A GUIDE TO

Scientific Writing

2ND EDITION

David Lindsay

 LONGMAN

Addison Wesley Longman Australia Pty Limited
95 Coventry Street
South Melbourne 3205 Australia

Offices in Sydney, Brisbane and Perth, and associated companies throughout the world.

First published 1984
Reprinted 1987, 1988, 1989 (twice), 1990, 1991, 1992, 1993, 1994
Second edition 1995
Reprinted 1996, 1997

Designed by Lauren Statham, Alice Graphics
Set in 12/13 Garamond 3
Printed in Malaysia through Longman Malaysia, GPS

National Library of Australia
Cataloguing-in-Publication data

Lindsay, D. R.

A guide to scientific writing.
2nd ed.
Includes index.
ISBN 0 582 803128.

1. Technical writing. I. Title

808.0665

The
publisher's
policy is to use
**paper manufactured
from sustainable forests**

Contents

Chapter 2   Second Draft—Getting it Together

Chapter 3   Third Draft—Readability

Chapter 4 Final Draft—Editing

Part II   Broadening the Scope of Scientific Writing

Chapter 5   The Review

Chapter 6   The Seminar or Conference Paper

Chapter 7   Student Practicals and Reviews

Chapter 8 *The Thesis*

Chapter 9 *Writing Science for Non-Scientists*

Preface
to Second Edition

THE MAIN PURPOSE of this book is to demonstrate ways of assembling facts, ideas and arguments in a form that is economical and enjoyable. Scientists' desks these days are crowded with a lot of dull, dry material essential to their work, but in no way making pursuit of knowledge easy. When they go to seminars they have to sift through hours of boring details to glean only a few minutes' worth of the information they seek. Occasionally into their drab world comes a gem of a paper that is a pleasure to read, or they are treated to an exposé at a conference that unfolds a clear message, unforgettable in the mass of mediocrity. By analysing what makes some papers, reports or talks enjoyable and what makes the others heavy going we can extract ideas that can put us on the way to making our work interesting to others. This book aims to analyse the construction of good papers and present those ideas that enable rapid and clear communication between writer and reader.

This book emphasises ways of assembling and constructing your material rather than the style in which it is written. There are many articles and books on style, from those put out by journals to tell you whether 'hour' should be abbreviated to 'h', 'h.', 'hr.' to those more profound texts such as Brogan,[1] Damerst,[2] and others that present positive ways of writing good and forceful English. Good style is an essential ingredient of a scientific paper but, as we shall see, other aspects are also crucial.

You may get the impression that my suggestion of introducing enjoyment into scientific communication means that I am advocating frivolity or charlatanism. Far from it. Science strives for precision and accuracy and you will see that I am advocating precision and accuracy in writing. Ideally, a reader or listener should be able to follow a piece of scientific thinking without backtracking, without distraction and without falling asleep. The only way to ensure that they can do so is to be precise and accurate. To be precise and accurate requires discipline—throwing away irrelevant material, setting straight things that are ambiguous and, above all, keeping expressions simple and in sequence. In short, I am suggesting that to write a good article is to tell a story accurately and well.

Experimentation is a sort of detective work with all the excitement of developing theories, sifting clues, reshaping theories and, occasionally, coming up with an answer. Detective stories, especially true ones, are among the most fascinating in modern literature. Why should scientific stories be any less entertaining?

All scientists are short of time and yet all are compelled to keep up with the literature. Their reading is selective and critical. They scan a lot of journals but don't actually read many articles in them. If you write articles that have the same appeal as five pages of the telephone directory, your name and your work will remain unread, unquoted, and unimportant. Some writers are still naive enough, or even arrogant enough, to believe that their fellow scientists are prepared to waste time on uninteresting catalogues of material in a quest for small pieces of information that may or may not be there. This assumption is untrue and is becoming more obviously so, as the volume of the world's scientific literature mounts exponentially.

Most of the ideas in this book are not my own but those gleaned directly or indirectly from the literature and particularly from thoughtful colleagues during many hours of debate. I am grateful for the advice and suggestions made by Brigid Ballard, Marcus Blacklow, Alan Robson, and by Reg Moir whose seminar is reproduced later in this book. I am fortunate in having a number of French colleagues, especially Dr M. Courot and Dr M. Bosc, whose concern for good writing is heightened by their obligation to read it as a foreign language.

Notes

[1] Brogan, John A. (1973), *Clear Technical Writing*, McGraw-Hill, New York.
[2] Damerst, William A. (1972), *Clear Technical Reports*, Harcourt Brace Jovanovich, New York.

Introduction

'If you haven't written it, you haven't done it'.

MANY SCIENTISTS and students of science find it hard to come to terms with this blunt, but absolutely correct, statement. After all, they have trained for many years to learn the scientific skills that have made them good scientists and yet very few of them have had any formal training in the specific requirements of writing about the scientific information that they are able to create. So, writing is a dull, difficult, relatively unfamiliar and occasionally embarrassing chore to be fitted into breaks in the more exciting job of discovering new things. Modern pressures to publish or perish have changed this view, although they have not necessarily improved the enthusiasm of many scientists about putting pen to paper or the quality of what they put there.

This book aims to develop principles of scientific writing that give it a logical base that should appeal to scientists. In particular, it endeavours to encourage writers to be efficient and logical in their use of words and to ensure that the purpose of each component of the written article is understood and achieved.

Scientists around the world communicate their findings and their reasoning through the 'scientific articles' which appear in one or other of thousands of journals published especially for that purpose. The first part of this book deals in detail with the development, construction and publication of the scientific article, the cornerstone of modern scientific communication.

The second half of the book uses the scientific article as a base to develop an approach to other written forms that scientists and students use to communicate their work. The thesis, the review, the seminar article, practical reports by science students and writing of science for readers who are not scientists, are all founded on the principles that underlie good science and skilful communication. Even if you are seeking information solely about one or more of these written forms I urge you to spend time reading the first part of the book to become familiar with the principles it advocates.

Part I

The Journal Article

Creating a Masterpiece in Four Drafts

1

First Draft—Getting Started

THE LENGTH OF AN ARTICLE is normally between 2000 to 10 000 words but you can write them only one at a time. This makes it difficult to envisage what the last word will be when you are concentrating on writing the first. Equally, it is difficult to remember the beginning of the paper when you are absorbed in the conclusion. Clearly, to write a good paper by progressing step by step from the title to the final reference is a task beyond the powers of concentration of most of us. The sheer mass of the task before us when we sit down to write the first word is often so daunting that we may feel unable to start. Fortunately, there are a number of features of scientific writing which you can take advantage of to make writing manageable.

The first of these features is the content of a scientific article. It is no more and no less than an objective and accurate account of a piece of research you did. It should not be designed to teach or to provide general background. It is simply about your own work and only calls on the work of others to substantiate arguments you might wish to make about your own work. If you recognise this early it will help you to clear up many of your problems as to what should or should not be included.

The second feature is the style of writing. Let us be clear about our objectives. We are about to embark on a piece of scientific writing, not a piece of English literature. If we look on the task of reporting experimental findings as a technique rather than an art, we can dispense with a common obstacle to getting started—an inability, real or supposed, to write 'perfect' English. Our paper should contain three ingredients, precise logical science, clear and concise English, and the idiosyncrasies of style demanded by the journal to which it will be submitted. Not all of these will be achieved in one attempt. Instead, we should plan to write a series of drafts, each improving on the one before, and each concentrating on a different ingredient. For example, four drafts might concentrate, in order, on getting your ideas on paper; the scientific content and the logic; the fluency of the English expression; and the style of the journal. You may even feel happier preparing more than one draft for each of these components.

In the first draft the task should be to develop a series of logical scientific statements. Of course, there is nothing wrong with writing good English or complying with the style of the journal at this stage. I am merely advocating that you do not let them hold up your progress. Once you have decided on what you wish to write you should write it. If English is not your native language you may even wish to write difficult passages in your own language. The point about making a draft is that it gives you something to build on. Twenty sentences of doubtful literary merit that have got you deeply involved in the paper and enthusiastic to continue are infinitely better than one magnificently composed sentence that has left you exhausted. If your ideas are allowed to flow freely, your writing will usually do the same. Blemishes within sentences, which, after all, can be easily repaired later, will be compensated for by a coherence between sentences.

Non-native speakers of English, in particular, may feel obliged to have their final draft corrected by an expert in English. If this is the case, make sure that it is the final draft and not an earlier one. Get the Science right first. Correcting grammar, sentence structure and idiom is relatively simple. Correcting illogical flows of ideas, and poor 'science' is impossible unless the English expert is also a scientist familiar with your field.

One of the most convenient features of scientific articles is that they are divided into clearly delineated sections. This is helpful because at the stage of the first draft we only have to concentrate on one section at a time. We can thus visualise more or less completely the whole section while we are working on any part of it. If our minds cannot encompass in one quick pass all the thoughts we want to incorporate in each segment, then we should further divide the paper into smaller, more manageable segments. Once all the segments have been written, the paper can be constructed by piecing them together. A preliminary draft in such a form is comparatively easy to amend and alter.

For the average paper the divisions normally acceptable to, and, indeed, demanded by the editors of journals provide sufficient segmentation for this purpose. Usually these are:

- Title
- Introduction (Why did you start?)
- Materials and Methods (What did you do?)
- Results (What did you find?)
- Discussion (What does it mean?)
- Acknowledgements
- References
- Summary

The sequence given is practically invariable, except that in some journals the Summary may be at the end rather than at the beginning. That is not to say that you should be similarly constrained and construct your paper in a strict sequence. What is important is to start writing somewhere and build a feeling for the paper. Many people break the ice by drafting the Materials and Methods section because it is the most straightforward. Others experiment with ways of presenting the Results. We shall see in a moment, though, that whatever order is chosen it is helpful to attempt the Introduction relatively early on. Apart from this there is no rule; the sequence in your first draft depends on your data and how you wish to treat them. In many cases, with well-planned experiments you will already have the outline of some sections. You may decide to press ahead by expanding these before attempting the other sections.

For convenience only, we will deal with the sections here in the order generally used by journals.

The Title

The primary aim of writing a paper is to have it read. The title is the first—and possibly the last—a reader will see of your paper. Its importance cannot be overemphasised. You must try to produce something that is not only factual but stands out from the mass of other titles in the Contents section of the journal or in the lists of an abstracting publication. Literature-scanning services these days use a 'key words' system. You should attempt to pull out the key words from your material and try to incorporate them in the title. This has two advantages. It ensures that scanning services catalogue your paper properly, and it means that you will probably have the most descriptive title possible for your paper.

Look at the Contents section of the nearest journal you can find. See how many titles take the stereotyped form:

'The effect of [or 'The influence of'] . . . A on B'.

You will notice immediately how little incentive these titles give you to turn spontaneously to the articles to find out more. Such titles also relegate the key words 'A' and 'B' to the end where they risk being lost in computer-based search systems. Worse still, they do not tell you what happened. 'A' may have affected (or influenced) 'B' by making it better, worse, or it may not have changed it at all. What an anticlimax to read a paper in which the title announces that something is supposed to influence something else but the text shows that there has been no effect.

It is a good plan to begin the title with a key word. Wherever possible, put the most important word in your title in the position of power—the beginning. Then endeavour to hint at the main outcome of the experiment along the lines 'A improves [or 'reduces'] B'. For example, a title 'The influence of manganese on petunia leaves' conveys much less and is less attractive than 'Manganese brightens the colour of petunia leaves'. Consider the title, 'The influence of time of lambing on the performance of Merino ewes'. Two questions spring to mind, 'What sort of influence—good or bad?', and 'What performance—running, jumping, producing wool, producing milk . . .?' A better title is 'Merino ewes produce more wool if they lamb in spring instead of autumn'. We have added one extra word but have increased the information and the impact disproportionately.

Some titles appear to have been designed to put readers off. Those that start 'Some aspects of . . . ', 'Observations on . . . ' give the impression that even if we read the paper we will only get half a story.

Titles should be positive, brief and specific.

As an exercise, look for some dull titles in a familiar journal and for each title read the summary of the paper that follows. Using this information compose a new title that is at least as brief, more specific and informative, and more helpful to a 'key word' system than the original. Such an exercise is not only good practice but it will quickly illustrate the poor standard of the headings used.

The Introduction

The purpose of a scientific paper is to communicate new scientific information to scientists. Its first objective is therefore to demonstrate that the story being told is worth telling. Here, in the Introduction, is the opportunity to set up the paper as a clear example of scientific thinking. The Introduction goes much further than just stating the problem and orientating the reader towards the relevant literature. It should guide the reader gently but precisely through a piece of logical thinking that ends in a statement of what the experiment is about and what alternatives might be expected from it. If the Introduction has done its job well the reader will no longer be a passive recipient of your information but an enthusiastic searcher.

If you examine some good papers carefully you will see that most of them introduce the paper with two paragraphs or, if it is a short paper, with a single paragraph containing two distinct sections. The second part is shorter than the first and contains the key to all good scientific papers, the hypothesis. The first part has no other purpose than to present the hypothesis as a reasonable scientific proposal.

The hypothesis

Beveridge[1] calls the hypothesis 'the principal intellectual instrument in research'. The hypothesis is so central to the so-called scientific method and, because of this, to the paper we are about to write, that it is worth spending a little time examining it.

There are several texts on the philosophy of science and scientific method that deal extensively with the hypothesis but, in short, we can describe it as 'a reasonable scientific proposal'. It is not a statement of fact but a statement that takes us just beyond known information and anticipates the next logical step in a sequence of supportable statements. It has to have two attributes to be useful in scientific investigation: it must fit the known information and it must be testable. To comply with the first attribute you must have read the literature in your field. To comply with the second, you have to do an experiment. The paper you are about to write concerns nothing other than these two things. You can see why the hypothesis is so central to your writing.

Of the infinite number of things you could have investigated in the world and the infinite number of ways of doing the investigation you chose only one or two. Your job in the Introduction is to convince the reader that you have chosen sensibly on both counts.

As a simple example, suppose you have taken a series of measurements of the concentration of iron in the leaves of groups of orange trees which have been fertilised with different amounts of lime. 'Why on earth would you want to do that?', the reader asks. 'Because your boss told you to do it? Because your laboratory has just bought a new machine that can measure the concentration of iron easily?' Let us supply the reader with some information.

1 Orange trees in your area have been showing signs of iron deficiency in the last few years.
2 In previous years yield and quality of fruit have been improved by dressing the soil with lime.
3 Lime raises the pH of soil.
4 Uptake of iron by plants is low at high pH.

These statements put some sense into your activities and certainly show that you are working in an area that needs sorting out. But do they explain enough? From numbers 3 and 4 it seems clear that you are going to get lower concentrations of iron in the leaves of trees given lots of lime, so why bother repeating someone else's experiment?

So far we have given:

1 the objective of the experiment
2 the pertinent information

but we haven't put in the important ingredient—the hypothesis.

Let us recapitulate. Orange trees in our area need lime and they need iron. These two are apparently beginning to antagonise one another. But maybe there is a level of lime which will supply sufficient calcium and at the same time allow iron to be absorbed in sufficient quantities to grow good oranges. Suddenly we have an hypothesis; it seems to fit the information that we have on the subject, and it is testable. Now we know what the experiment is all about and why it is being done the way it is.

Now we are ready to write the main paragraph describing the experiment as testing the hypothesis that, in the soils of Rattlesnake Valley, there is an optimum level of application of lime that will provide sufficient calcium for orange trees without creating a deficiency of iron.

Readers know what you, the writer, are looking for and can assess every sentence of the rest of the paper against these expectations. Instead of thrashing about trying to find out what it was all about, they will now read ahead in anticipation. For this reason there is no other single statement in your paper as important as the hypothesis. Having written it you should write it out in red ink, in capital letters, or whatever method will emphasise it most, and pin it up in front of you. You, the writer, and the reader are now both fixed on the same objective but the reader will probably finish the paper in ten minutes and is unlikely to forget the objective. It may take you several weeks or months to write your paper so a constant reminder of your hypothesis will serve to keep you on the right path.

In our example we finally presented the hypothesis in the form 'to test the hypothesis that . . . ' It is possible to present an hypothesis in many ways such as: 'If A . . . then . . . ', or 'If A, B and C . . . , then X . . . ' In short, it must be in any form that presents a reasonable expectation of the results based on the known information. What we must avoid is a statement which says something was attempted in a certain way to 'see what happened'; or 'it seemed of interest to examine this phenomenon further'; or 'There are no reports in the literature of a study of this, so one is presented here'. All of these betray a randomness of thought and a lack of scientific discipline that signal that the next few pages of text are going to be hard to read. With these sorts of statements you can, and probably will, present and discuss almost any results and observations that come to your mind so long as they conform to the general area you have so vaguely defined.

The reasoning behind the hypothesis

With the last part of the Introduction concluded, the first part can now be filled in. Many people have difficulty in deciding what should go into the Introduction, what work should be quoted and what should be left to the Discussion or even left out. The decision is easy. Only that material forming part of a logical series of statements leading to your hypothesis should be used. Just because you know of work in the general area, or because some well-known scientist might become annoyed if they are not quoted is no reason for having them in the Introduction. You are not in the business of doing favours for other authors by slipping their names in somewhere. Only if they contribute to the development of your logic should they be included.

Far from being a loose preamble the Introduction becomes a very tight, clearly defined piece of writing the moment that you settle on the final form of your hypothesis.

The hypothesis in its final form should have been developed by either inductive or deductive reasoning. You should use the opening paragraph to make clear the inductive or deductive logic that makes it a sensible proposition to test.

Induction is the logical process of assembling facts until a conclusion, usually a generalisation, is reached.

For example, take the following pieces of information:

1 Hay fed to dairy cows results in a low concentration of propionate in the blood, and milk with a high concentration of butterfat.
2 Concentrated feed results in high concentrations of propionate in the blood, and milk with a low concentration of butterfat.
3 Lush green pastures result in milk with a low concentration of butterfat.

These three pieces of information, when related, suggest that green pastures may produce high levels of propionate in the blood. This is an induction. It is a very short step to convert this to an hypothesis that 'lush green pastures increase the propionate concentration in the blood', which can be tested experimentally.

Deduction is the application of an accepted law to a specific situation. If from the above three statements we accept that hay produces high butterfat, we can deduce that if we feed hay to cows on lush pasture we can raise the concentration of butterfat in the milk. Again the hypothesis to be tested almost writes itself, 'Supplements of hay will raise the butterfat content of milk from cows grazing lush pasture'.

In setting up the hypothesis it is well to remember that a result that supports your hypothesis does not mean that the hypothesis is infallible. For example, if we set up an experiment to test a generalisation and find that our results fit the hypothesis this is no more than additional evidence that it might be right. Another series of observations made under slightly different circumstances might fall outside the generalisation you have made.

On the other hand, if your observations cause you to reject the hypothesis (and your experimental methods are sound) then you can be much more positive about your conclusion. For this reason, and for the sake of the written story, it is often preferable to frame a proposition about the known information in a way that your results may reject.

For almost eighty years, and after countless experiments, it has been accepted that female sheep shed more ova and produce more twins if they are placed on a 'rising plane of nutrition' six weeks before they are mated. The hypothesis that says that 'the heavier the sheep becomes the more eggs it sheds' has been tested and accepted so many times that it has become an acknowledged law. As a result, farmers seeking more twin lambs fed the best feed possible to the ewes for six to eight weeks to make them gain weight before joining the rams. Several years ago, C.M. Oldham had the idea that perhaps weight and the production of twin lambs were not necessarily associated in a cause-and-effect relationship. By means of an endoscope, he was able to examine the ovaries of ewes daily after the beginning of supplementary feeding and found that the enhanced capacity to produce twins appeared as soon as six days after feeding—long before the animals had the chance to gain weight. Nobody before him had thought to test the animals for fecundity until after they had gained weight. In presenting his experiment, Oldham was able to refer to the data that supported the relationship between gaining weight and twinning, and then present this as a testable hypothesis. His data enabled him to reject the hypothesis which, as a consequence, had to be modified after almost eighty years.

Is an hypothesis obligatory?

Not all papers report the results of experiments to test hypotheses. Some present the results of surveys of new material or new areas. Some report observations on populations of people, plants, animals, landscapes or compounds to which no treatments were applied. How do you make an hypothesis out of such work?

Easily. It is difficult to believe that in this day and age anyone can afford the luxury of wandering about collecting and assembling data at random. There has to be a reason for your doing what you did and you must have had some expectation of what you might find. An expectation of what you might find is an hypothesis! No matter how tenuous the reasons for your expectations, the thinking behind them is the background your reader needs to assess your results. Even if you were wide of the mark but your results are still interesting, he has a thread to follow while reading the paper. For example, you might have gone into the desert to measure the insect population on the native plants. There won't be much interest in a paper which is introduced by no more information than that. What reason could you have had for doing such work?

- Perhaps you believed that the insects you found would help fill a gap in a taxonomic sequence?
- Perhaps a new irrigation scheme is to start and you wished to find out the likely parasites and predators of crop plants in the native population?
- Perhaps you believed that some economically valuable insects could be collected and used elsewhere?

Each of these is a possible hypothesis, and will give purpose and direction to subject matter which might otherwise resemble the index of a street directory.

I have stressed the necessity to arrange your arguments and your hypotheses in a logical, precise order. Everyone knows that even brilliant scientists (in fact, especially brilliant scientists) do not always think in this way. Many of the great discoveries of science have developed from flashes of brilliance which often came under unusual circumstances. Newton was supposed to have developed his laws of gravity after being struck by an apple while relaxing under an apple tree. His brilliance lay in being able to relate a 'happy accident' to the known facts and then build up the laws of gravity. Fleming had his stroke of luck when, in sub-standard working conditions, a plate of culture medium became contaminated. He too had the brilliance to think through the consequences of this, until he and Florey isolated penicillin. Parkes and Polge discovered that germ cells can be protected from freezing temperatures, because a technician made a mistake and mixed glycerol with some samples of semen which they were attempting to preserve. These scientists also worked carefully in the reverse direction—from the result to the reason for it—to achieve a major scientific breakthrough.

Down at our level, many of our ideas also come from inspiration. Fortunately, there is no way that we can train ourselves to develop ideas logically. If so, we could leave the whole of scientific discovery to computers. Most ideas do not stand up to testing against the facts we find in the literature or through experimentation. Sometimes, we set up an experiment to test what seems to be a good hypothesis at the time, only to find that our techniques are inadequate or our ideas were not so smart after all. It is not rare for such experiments to yield interesting data that could provide information for an hypothesis different from the one we were originally testing.

If the processes of scientific thought are so haphazard why am I asking you to set them out so logically when writing a paper? A basic rule of science, after all, is that we should be scrupulously honest. Shouldn't we record our ideas, discoveries and failures in chronological order? If we tested an hypothesis that we have now scrapped but in so doing saw how the results spread new light on a different hypothesis, shouldn't we say so?

You have had anything from six months to, perhaps, twenty years to test, reject, re-form and re-reject ideas and hypotheses. You have slept, eaten and worked with them and in the end you have come up with what seems an important piece of information. Your readers have about a minute to cover the same ground. You are therefore obliged to give them only the distilled essence of your thought processes.

Doing scientific research, and writing about it afterwards are not the same thing. They have very different objectives. Research is the finding of new information by testing hypotheses, rejecting or accepting them, refining and re-testing them, and finally coming up with new knowledge. Writing is the dispassionate recording of the knowledge in a manner that presents the data in an honest, plausible and straightforward way. The blind alleys you traversed, the disappointments, and the poor techniques along the way cannot be allowed to impair the reader's chances of seeing your new information as a clearcut piece of reasoning. Therefore, you should present only the hypothesis (or hypotheses) that you intend to test in your paper and you should present only that supporting information which makes the hypothesis sensible.

The Materials and Methods

This is the most straightforward section of the paper to write but it can get out of hand if you try to include too much detail. The skill in writing a good Materials and Methods section is in knowing what can be left out. Take a hard look at what you could describe and see what can be cut out without reducing the meaning of the paper. The criterion for a well-written Materials and Methods section is that a reasonably knowledgeable colleague could repeat your experiment after reading the description. The best way of checking is to find a knowledgeable colleague and ask him or her if your effort meets this criterion. If you are describing a technique with which you are almost contemptuously familiar you run a high risk of leaving out important details quite unconsciously. The view of an outsider can prevent this happening.

What can be left out?

Where novel techniques, or new modifications to old techniques, have been used they must be described fully and exactly. If the techniques have already been described fully, then it is adequate and desirable to refer to the paper where the technique was first (or best) described. Be careful though, to give credit where it is due. To refer to a recent paper in which the technique was used rather than to an older one in which it was originally described, is not simply discourteous to the original author but it fails to put the technique into historical perspective and it does not enable a reader to proceed directly to a description of it.

Details of factors like climate, soil types and live weights of animals are essential if they are what the paper is about, or are to be used in interpreting the results. If not, they should not be included. Data of this sort can kill a paper by drowning it in irrelevancies.

It is often appropriate to describe statistical techniques. For some reason statistics often appear to receive preferential treatment in that the most obvious manipulations are described in great detail. Statistical analyses, like chemical analyses, are normally the research worker's tools of trade, not the finished product. If you merely carried out a standard procedure like an analysis of variance, 't' test, or chi square, then simply say so. If the technique is more 'off beat' but well described in a published paper or standard text then a reference to the source will be sufficient. Only if you have performed some original mathematical gymnastics do you need to describe them and, even then, a reference or two can often shorten the task.

Sequence of headings

There is no set sequence for the Materials and Methods section, but if you jot down all that you believe should be included a sensible sequence will often become apparent. Sub-headings help to make a clear layout. There is another good reason for having a series of sub-headings in the Materials and Methods. This is probably the most skimmed over section of a scientific article, at least during the first reading. Good sub-headings ensure that the skimming is easy and fruitful. The sub-headings will describe sections dealing with the physical facilities and equipment, and chemical and statistical analyses. Normally, there is a heading Experimental or Experimental Procedure which shows details of the experimental plan. In other words it is describing the 'methods' part of the Materials and Methods. Almost all of the rest of this section will describe the 'materials'.

With the reader in mind you should also endeavour to put your Experimental Procedure or Design of Experiment section as early as practicable in the sequence. It is unfortunate that the sequence in the title for this section is Materials and Methods and not Methods and Materials. To describe materials, analytical techniques, sampling techniques and sources of information and chemicals before you describe what you are actually using all of this for—the methods—will inevitably result in the reader having to go back over the section to appreciate it. If you describe the way the experiment was done first then the reader will be able to see where and why the detailed sections about materials and techniques are included.

Four examples of well-chosen headings are given below which give you a reasonable idea of the experiment. The rest only adds detail.

EXAMPLE 1:

    Experimental Design

    Experimental Animals and Diet

    Treatment Infusions

    Sampling

    Analytical Methods

    Statistical Analysis

EXAMPLE 2:

    Design of the questionnaire

    Selection of individual recipients

    Follow-up questioning

    Analysis of data

EXAMPLE 3:

    Treatments

    Site

    Sowing the Plots

    Defoliation

    Sampling

EXAMPLE 4:

    Source and preparation of tissue material

    Immunochemistry

    Chemicals

Preparation of RNA
Preparation of probe DNAs
Analysis by nuclease S1 protection
Statistical analysis

The Results

Not only should there be nothing but results in the Results section but *all* results that you intend presenting in the paper should be in the Results section. No results should appear for the first time in the Discussion or, even worse, in the Summary. To do so creates extraordinary chaos in the process of comprehension. Readers either think that you have kept something from them or that they have missed something at the first reading. In either case they become justifiably annoyed, and their appreciation of the article is diminished.

In descriptive papers there is sometimes little or nothing to discuss after the results have been presented. In these cases, it may be sensible to add whatever discussion is necessary wherever it is appropriate in the Results and the section becomes Results and Discussion. Some journals allow this, but the practice is becoming less common. In fact, if the Introduction has been properly constructed and the reasons for doing the experiment are clearly stated, then you are compelled to discuss how the results met your expectations. It is usually neater and simpler to describe your results and then to discuss them in a separate section. A mixture of Results and Discussion in all but the shortest of papers invites chaos in the flow of arguments.

Another strong argument for separating Results from Discussion is that it preserves the objectivity of the Results. The purpose of the section is to present all of your results clearly and without comment. Readers are invited to draw their own conclusions which they will, no doubt, compare later with yours when they reach the Discussion section. To qualify each result, or group of results, with comments and comparisons gives the strong impression that you are trying to influence the objective judgement of the reader.

Despite the desirability of physically separating Results and Discussion, it is essential to have one in mind when composing the other. When sitting in front of notebooks and work books full of results from your experiment and wondering how to get them into a presentable form it is a good idea to remember that, in themselves, your results are not the most important new knowledge you are presenting to the world; you will be remembered and quoted for your interpretation of the results. In biological disciplines, if you, or anyone else, were to repeat the experiment you would not expect the treatments or observations to yield exactly the same numbers, but you would expect that the new data could be interpreted in the same way. I say this to encourage you not to lay before the reader large helpings of figures in endless rows and columns. Instead, you should present readers with carefully chosen and distilled information that best enables them to understand your interpretation which will follow in the Discussion. If you have been conscientious in the collection of the data during the course of your experiment, you may find that you have material to occupy the space of two or three papers. The successful construction of the Results depends on your choice of the material to present, and your decisions on how to present it.

What to present

The hypothesis in the Introduction lights the way to selecting the material for presentation. The hypothesis says what the paper is about, and results that do not relate to your hypothesis should not be included. If you decide that there is an important exception to this, it must be given very little space or you will find that your paper begins to lose direction. When you start making exceptions, it is often time to take stock of the results you have. Perhaps, if you have a lot of material that is unrelated to your hypothesis you should think about writing another paper. Alternatively, you may decide to abandon the paper in its present form and present the material under a different hypothesis.

Sometimes this selection process means ruthlessly culling results that you worked hard to collect. Readers are not interested in your capacity for work. In fact, if you overwhelm them with data they may be more impressed with your lack of discrimination. Regard heavy pruning as a normal part of paper writing.

How to present it

You still have to present the selected data logically and concisely. Most data require some treatment. This may vary from complicated statistical analysis to simple tabulation of results and the calculation of a few means. It is always worthwhile attempting some alternative methods of treatment and presentation before deciding on the most suitable. At this stage you should be forming the arguments which will become the backbone of your discussion. This, too, is a process of trial and error to find the best alternatives for final presentation. Remember, you will be referring back to certain highlights of your Results for the basis of the Discussion. So it is essential that in the Results section the important points are highlighted. Whatever your final presentation—graphs, tables or text—you should arrange it so that the key information and key figures are in prominent positions.

What form of presentation? Tables, figures or text?

When more than a few numbers are involved it is preferable to use tables than to attempt to display the numbers in a horizontal line of text. But remember that using tables and figures does not absolve you from the responsibility of making the text a coherent story. A useful rule is that the text should be readable without the tables, and vice versa. This does not mean that the text should present the same data as the tables. The text gives you the opportunity to reinforce those aspects of the tables that will be particularly important when you come to the Discussion. It is highly unlikely that every figure in every row and column of a table is as important as every other one. In the text you guide the reader by drawing attention to important parts of the table. You can make the text less sterile by saying 'Table 1 shows that moonbeams have more emissions in the blue spectrum than the red' rather than 'Results are shown in Table 1'.

To make the tables self-supporting, it is important to use full, descriptive titles and to make the row and column headings informative. A title for a table that just says 'Milk production of treated cows' is quite inadequate. The title must designate the number of the table, and it should give the essential details of what it describes. Thus a title 'Table 4: Milk production in litres/day of 10 Jersey cows during the first 30 days of lactation after injection with 10 mg of iodinated casein' is far more acceptable.

Row and column headings should also be complete. There are few things more annoying than a table in which the descriptive headings have been replaced with indecipherable codes. Apart from being readable these headings should always specify the units, for example, gm/day, ml/100 mg, or %. The body of the table should show numbers only and not become cluttered with the units. If all the units in a table are the same, this information can be given in the title as was done in the previous example. If there are any omissions or abnormalities, they should be explained in footnotes. Incidentally, tables and figures are the only place in scientific articles where footnotes are acceptable. Footnotes should also be used to explain abbreviations, symbols, references, and statistical information, even if these explanations are also given in the text. A footnote that invites the reader to 'see text for details of treatments' implies that the reader is clever enough to read the text and the table simultaneously. A good plan is to ask colleagues if they understand fully what your table is about without referring to the text. If not, you must clarify the table by adding sufficient material to either the title, the headings, or the footnote.

Graphs or figures are often thought to carry more impact than tables, especially where continuous changes associated with continuous inputs of treatments are being described. Even so, it is sometimes difficult to decide which is the more appropriate. On the face of it, graphs that are simple are usually easier to digest than a group of numbers but are far less precise. If the aim is to use the material to show *qualitative* features of the data and gross differences, graphs are ideal. If the testing of the hypothesis calls for a close, *quantitative* analysis of the results, then a table containing the exact numbers is a better presentation. For example, if we wished to show that wool production of sheep increases with increasing concentration of protein in the diet up to fifteen per cent and is not stimulated further by concentrations higher than fifteen per cent, the story can be simply and completely told by an asymptotic graph in which the horizontal axis shows the concentration of protein. In this example, where we are more concerned with trends than with absolute numbers, the scale on the axis is of little importance. The purpose of the graph is to simplify and to make the data more 'graphic'. On the other hand, graphs are almost useless when a detailed analysis is important. If we wished to demonstrate that the daily requirement for wool production in a sheep is 1.7 g nitrogen per g wool, the precise numbers from which this estimate was derived are essential and a well laid out table is the appropriate medium. Graphs are showy but they do not allow you to summarise the results as tables do.

However, before you opt for a graph and dismiss tables as boring substitutes, consider the views of A.S.C. Ehrenberg who is not only a strong advocate for tables but has presented an approach to constructing them that supports his opinion.[2]

According to Ehrenberg, tables are not just rows and columns of numbers. Good tables should present the numbers in a way that highlights the patterns and exceptions in the data. By inserting marginal averages and siting them so that contrasts and comparisons are easy, you can make the reader aware of the major balance of the data at a glance. For example, if you decide that the discussion may involve the contrast between two particular means, those means should be so positioned in a table that they are close enough together to compare visually. If the object is to show the relationship between two or more series of numbers, arrange them in columns rather than rows. Reading down columns is much easier than reading across rows and the patterns emerge more quickly.

For an example, compare the following two tables on page 22. They both show the same data—the production of saw logs from seven forest sites over five years. Table 1 lists the sites logically in alphabetical order and faithfully presents the annual production to two decimal places, but it is far from user-friendly. Table 2, on the other hand, sets out to make these same data visually comprehensible by using at least four helpful techniques.

First, the writer has estimated that the extra information in the two decimal places shown in Table 1 is completely wasted when describing differences in the production of saw-logs and has rounded the precision to a manageable, but totally appropriate level.

Table 1   Yearly production (in '000 tons) of saw-logs from 7 forest leases

Lease	Year				
	1990	*1991*	*1992*	*1993*	*1994*
Cedar Junction	137.63	129.17	149.38	117.21	183.40
Dead Dog Hill	29.70	30.79	33.53	27.41	34.64
Heartbreak Hill	16.54	19.38	19.88	16.59	21.62
Millstream	142.63	137.60	171.79	162.40	194.26
Paradise	206.48	274.56	275.98	213.78	303.35
Queen's Ridge	47.32	51.83	53.73	49.10	60.23
Rapids Falls	63.54	77.82	81.76	54.20	89.49

Table 2   Yearly production (in '000 tons) of saw-logs from 7 forest leases

Lease	Year					
	1990	*1991*	*1992*	*1993*	*1994*	*Average*
Paradise	206	275	276	214	303	255
Millstream	143	138	172	162	194	162
Cedar Junction	138	129	149	117	183	143
Rapids Falls	64	78	82	54	89	73
Queen's Ridge	47	52	54	49	60	52
Dead Dog Hill	30	31	34	27	35	31
Heartbreak Hill	17	19	20	17	22	19
Average	92	103	112	91	127	105

Second, the visual impact of the relative productivity of the seven sites is emphasised by placing the sites in decreasing order of productivity. After all, there did not seem to be a lot of logic in having the sites in alphabetical order.

Third, a small but discernible gap has been added to emphasise the clear differences between the top three sites and the other four.

Fourth, row and column averages have been inserted to provide better orientation for the reader. Again, a visual gap has been left to distinguish the averages from the rows and columns of data. The relative productivity of the sites is confirmed by the row averages, and new information that illustrates the differences between years is added in the column averages. Note how quickly you can see the very high productivity in 1994. Ehrenberg emphasises the desirability of averages rather than totals at the ends of columns and rows because averages can be readily compared with values in the body of the table.

In short, Table 2 has deliberately made obvious the patterns and exceptions that the data have to offer. The table is telling you the results even before you read the numbers.

Just as data from tables should not be quoted verbatim in the text, histograms or graphs should not duplicate data already given in tables. Editors dislike wasting valuable space and money on such duplication and, even more importantly, readers quickly tire of being forced to read the same thing twice. Repetition should be reserved for oral presentations, as we shall see later.

Use of statistics in presentation of results

Statistical analysis is a powerful tool which allows you to place probabilities on your results. It prevents you from getting carried away with differences that could be due to chance and it gives support to statements that claim that treatments are effective. But, remember that the responses or the differences are the important things, not the statistical technique that has given you the confidence to claim them. You should ensure that levels of probability are clearly stated, but you do not need to present tables to describe how you derived the levels of probability. They are no more essential to your paper than intermediary chemical analyses may be to your final conclusions about chemical constituents.

There are good techniques that enable you to present the statistical information in the same tables or graphs you are using for presenting the data. For clarity, it is preferable to summarise large masses of data by reducing them to totals or averages. Averages are preferable to totals because they present the summarised data in the same scale as the individual values and this makes visual comparison much easier. If you do this, you can also indicate the degree of variation in your original data by presenting the standard error of means or the standard deviation of individual records. These are not the same thing, so you should not simply write '12.6 + 1.3' because it is not clear whether

1.3 is the standard error or standard deviation. Putting (SE) or (SD) in parenthesis behind the '12.6 + 1.3' can clear this up. Even when numbers in tables are arranged to allow a clear visual comparison between them an order of statistical significance should also be included. Without it, results cannot be reliably interpreted. The word 'significant' has other meanings than the statistical one but because in modern scientific literature 'significant' usually means 'statistically significant' it is a good idea to avoid using it in any other sense.

What is a 'good' number of decimal places to use, or how many significant digits should there be in recorded numbers?

There are at least three rules that can be applied to answer this question. The first is based on common sense, the second on statistics, and the third on readability. None of them covers all possibilities but all of them can be useful in helping to avoid 'number overload'.

Common sense says that you should not present absolute numbers or means at any greater level of precision than is measurable by the instruments you used. If you take individual recordings of data in whole numbers it is silly to present averages of those numbers to two or more decimal places simply because your calculator or computer was able to give them to you in that form. To do so is to display a gross lack of scientific discrimination and, worse still, a lack of respect for your reader. There are exceptions, of course, when the point of the scientific exercise is to question or to study very precise measurement, but this is very rare. For example if you were using the sex ratio of human births as part of a discussion on disease you would probably be happy with the ratio of 50 females to 50 males and this would be sufficient for your further arguments. However, if you happened to be discussing those rare and, possibly, sex-linked prenatal conditions that may affect male embryos and foetuses, but not females, you would need to work with a more exact ratio like 51.46 females to 48.54 males.

Statistically, Sokal and Rohlf[3] suggest that the last digit in a number should indicate the limits of the scale on which you believe the true measurement to lie. So, if you record the weight of a horse as 372 kg you are implying that it weighs between 371.5 and 372.5 kg. If, for a sample of a chemical substance, you record 0.0367 gm, then you are inferring the true weight to be between 0.03665 and 0.3675 gm.

In the case of the horse the unit increments between successive readings are assumed to be 1 kg. In the case of the chemical, the increments would be 0.001 gm . Sokal and Rohlf propose the rule that, where possible, the number of decimal places or the number of significant figures to record should be such that the number of unit increments between the smallest and largest reading should be between 30 and 300. To have more than 300 is being too precise for the data-set and to have less than 30 makes the data too prone to recording or rounding errors when moving from one reading to the next in sequence. So, if you were weighing school children ranging from 30 to 50 kg to the nearest half kilogram you would have 40 unit increments from smallest to largest, a satisfactory number. A precision of 0.1 kg would also be satisfactory, giving 200 unit increments.

The third view of data reduction based on the readability of the final information is the 'Ehrenberg Principle'. Ehrenberg concludes that our short-term memory which we use momentarily for recalling one number while we compare it with another is hindered and confused when we ask it to handle more than two significant digits at a time. If, on the other hand, we reduce our numbers to this level we are capable of amazing feats of numeracy, previously denied to us when we had to wrestle with big groups of digits. Take for example the numbers 473.64, 457.31 and 162.78. We can recognise instantly that one is smaller than the other two but we have to struggle to deduce much more. If we reduce the numbers to two significant digits, 470, 460 and 160 we see immediately that the first two are almost identical and the last is about three times smaller than the other two. In fact, if the numbers had not been reduced for us we would have tried to reduce them mentally anyway before attempting to make meaningful comparisons. Ehrenberg would argue that, in this case, the reduction in precision is between 0.8 and 1.7% which, for statistical purposes at least, is negligible. If a reduction in precision of this magnitude for these data on scientific grounds is also appropriate, then reduction is an excellent way to simplify interpretation. If that worries you, remember that you would have probably accepted an even greater loss of precision if you had decided to present the same data in a graph of some sort.

The Discussion

In a research paper Discussion is really short for 'Discussion of Your Results'. It is not a section in which you launch into a review of the literature on the subject. All literature cited must have the function of supporting arguments about your results.

No matter how simple an experiment, it needs to be interpreted and readers who have followed your paper so far will look with anticipation for your interpretation. If your Introduction and Results sections have been skilfully set up they may be beginning to make some interpretations of their own. If so, you are fortunate, because you need only guide them gently by logical steps to see things from your point of view. If you have already confused them you will have to use sledge hammer force to get your view across. Since they have the option of giving up and moving on to the next article in the journal your chances of doing this are very low.

Assembling the Discussion

Before getting down to deciding what to put in the Discussion re-read your hypothesis. Mentally match up its expectations with your principal results. The hypothesis is the uniting force throughout the paper but this role is never more important than in the Discussion. We have seen that in all of the sections so far the material that should be included is strictly circumscribed and you have had surprisingly little scope. Here, at last, in the Discussion is your chance to impress the reader with your ability to interpret and innovate—in short with your capacity as a scientist. Don't waste the opportunity by rambling.

While you were collecting, processing and tabulating your data, you will have formed a number of ideas that might be developed in the Discussion. These ideas which might come to you under the shower, on the bus or almost anywhere, need to be refined and related in a logical way to your data and to the literature. Many of them will perish in the process, and this is normal, but a few will come through as the important features of your discussion. We will call these developed ideas 'arguments', because you will have to justify them in the face of what is already known of the subject and you will have to present their limitations as honestly as you can. The Discussion then becomes a collection of arguments about the relevance, usefulness, and possibilities or limitations of your experiment, and its results.

Development of arguments

Each of your arguments can be considered as a separate piece of logical writing and will normally be the substance of a complete paragraph. The technique of developing arguments is identical to that of good paragraphing. A paragraph is the development of one idea. It gives the readers visual help, in breaking up the total bulk of the Discussion and helps them to absorb your points one at a time. A good paragraph has three components. Let us look at them and see how we can use them to develop our arguments.

1 Topic sentence.
2 Logical development.
3 Either a concluding message or, if a new, but related, idea is to follow in the next paragraph, a summary of the point reached in the development .

1 *Topic sentence.* Reading is much easier and more effective when we have some idea of what we are about to read. As writers therefore we must digress from our logical sequence a little by starting the paragraph with a mini-summary of what is to follow. This is often called the topic sentence. The topic sentence may in fact paraphrase the main point you wish to make in the paragraph. It immediately attracts attention and puts the reader on the right mental wave-length to receive the ideas on the logical developments which follow. As well as signalling the substance of what follows, the ideal topic sentence should also act as a link with the previous paragraph. This enhances the coherence of the whole discussion. Let us look at examples of some good topic sentences.

Corn plants sown before July are not only resistant to insect attack but they produce bigger grains . . .

By contrast with the daily rate of gain in live-weight, wool production is unaffected by high temperature . . .

Both of these meet the dual roles of referring to the previous arguments (dealing with resistance to insects, or daily gain in live weight) and opening up the new paragraph (which will discuss size of grains, or wool production).

2 *Logical development*. Having whetted the reader's appetite you can now get down to the business of using facts from your results and combining them with other facts or theories to make your point. Your aim is to draw a sound conclusion by deduction, induction or a mixture of both. For example, you may believe that your results allow you to make a generalisation not previously possible. This would be developed by assembling the essence of your results and, possibly other results from the literature, by the process of induction. You may also feel that your results have a certain new application, and your argument, to demonstrate this, will be based on deduction from your own and others' results. Each argument is unique and is supported by its own set of facts so there can be no specific rules here. However, there are a number of fallacies of logic which we must avoid and it is worthwhile examining these carefully.

a Generalisations: An invariable property of biological data is their variability. Thus premature generalisations based on a few preliminary results, or on results obtained under a limited set of conditions could lead to disaster later on. A valuable use of statistical analysis is to minimise the chances of making foolish generalisations. If the early evidence points your way, you have reason to be enthusiastic but do not let it tempt you into rash conclusions that put your logic into question. Even when statistical analysis shows that your results are unlikely to be due to chance, your generalisation must be guarded. The response of certain species of clover to potassium at a research station cannot be interpreted as showing that all species of clover should be fertilised with potassium throughout a whole agricultural region.

b Authority: There is no ultimate authority in science. Even Newton's laws of conservation of energy, which stood for centuries, were modified by Einstein, and Einstein himself is constantly being challenged. However, the cornerstones of most scientific arguments are one or more authoritative sources. It is impossible to go back to first principles in every case. So we have to accept certain concepts and statements as being acceptable, for the time being, as authoritative sources. So be careful that your choice of authority does not jeopardise your arguments. If your source of authority is out of date, or controversial, or simply wrong, your whole argument crumbles. In modern science, principles and concepts are being constantly revised in the light of new evidence. Your very paper may be presenting such evidence. You must be sure that the principle on which you are relying is currently accepted and recognised. If you have reservations about the authority you are quoting but can find no alternative, your reasoning should be appropriately modified to make this clear.

c Expressions of confidence: Your conclusions should be expressed according to the force of your data. If you have no conclusive evidence don't dither around with expressions such as 'It may be possible that . . . ' or (worse) 'The possibility exists that . . . ', which immediately suggests that you do not believe your own data. On the other hand, slight differences between treatment and control plots do not permit you to say 'There is a clear indication . . . ' or 'There was a marked response . . . '. In these cases it is safest if you do not develop your argument beyond giving the actual values. This is honest, factual, and eminently scientific.

3 *Concluding message.* To round off the argument and the paragraph you can emphasise in the final sentence the key point of what you have been developing. For example:

If grain must be larger than 3.5 g for processing then the crop must be planted before July.

There is thus no reason to believe that high summer temperatures will depress wool production.

Here we have examples of a specific, and a general, concluding sentence. Both of them have a clear message which they deliver emphatically. The reader will, of course, want to be convinced of the reasoning behind the conclusion that you have made. If the argument has been well developed it will be there in the body of the paragraph.

In practice it is a good idea to play with the sequences of information in each argument in the form of notes. In this way, you can decide finally on what sequence seems most logical and can therefore be understood most quickly by a reader.

Balancing arguments

As you begin assembling these arguments you will become aware that some are more important than others. It would be most unusual if you did not wish your reader to gain the same impression as you about the relative importance of your arguments. You will have to balance your arguments so that the important ones appear instantly important and their impact is not reduced by minor arguments.

At this point it is an excellent idea to set out in note form all the arguments that you expect to use in the Discussion and clarify for yourself the relative value of each. Examine each thoroughly and give it a grading. I use the gradings AAA, AA, A and B but any four symbols denoting descending order of importance will do. The partitioning is based on the following criteria:

AAA: Those arguments that are relevant to the original hypothesis but which allow you to make a positive statement of acceptance or rejection.

AA: Those arguments that are relevant to the original hypothesis but which for some reason are equivocal, or which lead you to suggest further experimentation or observation before acceptance or rejection.

A: Those arguments based on your results, not relevant to your original hypothesis but which you consider sufficiently new or interesting to be worthwhile including.

B: Those arguments based on your results, not relevant to your hypothesis and of marginal interest.

The next move is to cross out all arguments in category B or any that you could not easily classify. Those that remain are the basis of your Discussion and you have classified them in descending order of importance. You must now make certain that you present them to readers so that they also classify them in the same order of importance.

Achieving impact

I suppose that if you have invented a $10 atomic bomb or an elixir that prevents old age your statement to that effect will have all the impact you desire. Short of this, impact has to be achieved by more subtle means. You are well on the way if you can give readers the impression, by visual impact alone, that the piece of information they are reading, or are about to read, is important even before they have absorbed the contents. This is possible, and by so doing, you can encourage the reader to rank the priorities of your arguments in the same way as you have.

There are several possible techniques and you have to use one or more of them deliberately to convey the exact perspective you want.

1 Size

The reader automatically relates the length of text devoted to an argument to its importance. Newspaper editors use this technique on the front page, but they are able to increase the area and importance of a news item simply by using big type. You can't do this and, to make things worse, your most important point may be the most simple to develop. If it takes only a few lines, and a minor argument takes three-quarters of a page, your whole Discussion is visually, and probably logically, unbalanced. This does not mean that you should react by adding irrelevant sentences to your main item to increase its size relative to the minor argument. Rather, you should ensure that minor parts of the Discussion are dealt with in one or two sentences so that they do not get undue emphasis. Of course, it is unlikely that a strong argument will occupy only a few lines. It will probably have several implications and applications, each of which must be developed in the argument. If not, it is probably a sign that you should think twice about its ranking.

2 Position

Many writers think that the way to develop their discussion is to begin with the more trivial and unconvincing of their arguments, building to a climax in the last paragraph. It seems to them desirable, and possibly essential, to tie up as many loose ends as possible before coming to the main points. From the readers' point of view (and, in writing, theirs is the only point of view to consider) there is nothing more frustrating than to be presented with a battery of trivia when they are searching for main conclusions. Unless they have unlimited time, or an unusual interest in the work, they will read your first paragraph and assume that you have little to discuss. They will turn to the next paper without bothering to get as far as your important material.

Just as in the construction of the title you lead with a key word, and with each paragraph you open with a topic sentence to put the reader clearly in the picture, you should design your Discussion to lead off with your strongest arguments. In fact, one successful author once told me that a rule he regularly uses for starting his Discussion section is to 'blurt out unashamedly' what he thinks is his main conclusion and then set out to convert people to his point of view. A rule like this gives you the best chance of capturing interest and compelling readers to go on to the rest of your paper. Contrary to what some writers believe, your important paragraphs may ensure that your less important ones are also read, but your trivial arguments will seldom encourage readers to press on to your better ones.

3 State what is important

It does not hurt occasionally to state 'The most important aspect of these results is . . . ' Do not overdo this approach as readers quickly tire of hearing of the importance of your work. It is a technique to use when you feel that the size and position of the argument in the Discussion may be inadequate to convey the emphasis you want.

4 Sentence construction

The best way of getting a message across is to make sure that its meaning is clear the very first time it is read. The first step is to construct the paragraph so that its topic sentence signals precisely what is to follow. Following this, the sentences should be simple, readable and in logical sequence. Impact is lost if sentences are woolly, or flowery, or ambiguous. The criterion of clear writing should apply to every sentence throughout the paper of course, but it is especially vital here where you are striving to make your argument stand out.

Speculation in the Discussion

There is no more controversial aspect of scientific writing than that of speculation. Editors disagree among themselves and with authors about how much speculation they will permit. With some it is entirely forbidden. The word 'speculation' scrawled across a paragraph not only denotes its rejection but conveys the impression that the work is unworthy, unscientific and imprecise. I cannot accept this view because I believe that good speculation is the spice of science. It is the means by which new untested ideas can reach a wide audience. Also, it often stimulates readers to generate ideas of their own.

But let me qualify my stand here. When discussing the development of hypotheses, we saw that an hypothesis is an idea that fits the known information and which can be tested. It is, in fact, a piece of speculation. The most fastidious editors will accept it in the Introduction because it is to be tested and validated or rejected in the body of the paper almost immediately after it is proposed. Speculation in the Discussion is left untested and so invites criticism. I believe that if a piece of speculation is developed from the results of the experiment in the same way as an hypothesis and meets the same criteria as an hypothesis, it is not only acceptable in the Discussion, but desirable. The only valid grounds for rejection of a piece of speculation are that it does not fit the known information, including that described in the paper, or that it could not be tested using known technology. These constraints are sufficient to prevent undisciplined theorising but, within them, authors have the opportunity to raise new ideas, the rarest and most valuable of all scientific resources.

The length of the Discussion

One of the most common faults of the Discussion is that it is too long. And that usually makes it tedious to follow. Common reasons for discussions being tedious are the use of unnecessary references and unattached sentences. If each paragraph is carefully planned and the aim of each paragraph is kept clearly in mind when it is written there should not be a problem. Another source of excess material in the Discussion is data repeated verbatim from results. The Discussion is meant to be read in conjunction with the Results and repetition is seldom necessary. You can always refer to tables and figures already presented. It is very difficult to read fluently a series of numbers given in the text. In the Discussion it is preferable to generalise, make comparisons and draw conclusions. Where exact values are essential only one or two key numbers should be used to avoid cluttering the argument. If you find that you have trouble making sense without re-quoting large segments of the Results it is a sure sign that your results are inadequately presented. In fact, it is common to find that the layout of the Results may need to be modified when final details of the Discussion are being worked out. This to-and-fro adjustment may involve changing a table to a graph to emphasise a trend you want to highlight. It may mean altering the row and column headings to point up a contrast or a pattern between figures. It can also mean deleting a lot of unwanted data from a table so that the remainder can be used directly in discussion.

Citations in the Discussion

References have a very important role in scientific writing and their use and citation should reflect this. Statements like 'There is general agreement that . . . ' or 'The literature suggests that . . . ' without references will just not do. Every statement you make must be supported by your own results, the results of others, or an authoritative statement based on the results of others.

There is a common tendency to regard the references as something relatively unimportant and minor to the main text, but unless they are accurate the validity of the text can be ruined. They must be accurately cited. That is, names must be spelled accurately, data must be correct, and citations in the text must correspond completely with those in the Reference section.

They must also supply the correct information. It is not sufficient to put in a reference to a paper that had something to do with what you are talking about. Readers wishing to follow your arguments more thoroughly should be able to find exactly what they are looking for in the paper to which you have directed them. If the paper is not the one from which the original information came but merely one that used the original information to develop other arguments the reader has been tricked. The ready availability of photocopying machines has made the task of accurate referencing much easier. A good rule is to take photocopies, or obtain offprints, of all material being referenced. In this way references are available instantaneously to check that, for example, when you say Jones did something, he did in fact do it and, moreover, recorded it in the exact article and in the exact year that you have cited. Your memory can play tricks, and libraries, even when just down the corridor, are often too far away to check information straight away. A dossier containing copies of all the references you intend to cite avoids this common source of error.

References have many uses. They can be used as the ultimate authority on which to base arguments. They can be temporary authorities whose validity you intend to challenge, or they may be considered to be obviously wrong. It is possible to suggest the emphasis you wish to give a reference by the way you word the text. Examine these statements:

All aerobic bacteria are sensitive to umptomycin (Bloggs 1994).

The implication here is that this is an accepted concept. Bloggs was the first to present it, and you, the author, agree. This emphasis is characterised by the placement of the author's name and the date in parenthesis at the conclusion of the statement.

Bloggs (1994) found (or showed) that all aerobic bacteria are sensitive to umptomycin.

This implies that this is a less well-known concept, Bloggs deduced it, and you agree with him.

Bloggs (1994) claimed that all aerobic bacteria were sensitive to umptomycin.

In this case you imply that Bloggs went against general opinion with his claim and you are, for the time being, retaining an open mind on the subject. Note how the word 'claimed' and the use of the past tense indicate doubt and open up the possibility of a change of idea in the light of more recent work.

Subtle changes of emphasis such as these can establish clearly your position relative to that of the authorities you are quoting.

The Acknowledgements

Technical help and worthwhile advice that you have received from others should be acknowledged simply and without affectation. Bodies or individuals granting money which supported either the research or the authors of the paper should be thanked. You can consider this a matter of courtesy or a matter of survival but in either case don't overdo it. Expressions such as ' . . . without whose invaluable assistance this work could never have been possible'; 'Professor J. Blow's constant and untiring guidance throughout the course of this investigation . . . ' are hackneyed and unnecessary. Even worse, they may offend the very people you wish to thank. If Professor Blow helped in setting up some part of the work you should be specific.

Statements like that above imply that he was at your elbow the whole time. If your paper has flaws or if Professor Blow disagrees with parts of it, he will not thank you for associating him with the work.

The References

Every journal has a rigid format for the citation of references. Unfortunately, this format is not standard among journals and is one of the most frustrating aspects of writing papers. The final typing of the references should be made only after consulting very carefully a copy of the target journal or, if one is available, the instruction to authors.

Some research workers keep an extensive card index system on subjects in which they are interested. If you develop a system which is complete it is easy to withdraw the appropriate card each time you cite authors in the text. The cards can be shuffled easily into alphabetical order and the References section compiled directly from the cards. If, like me, you do not have a good index system, you will still save a lot of time if you write out a card for each reference as you cite it and rearrange the cards at the end. There are now computer databases configured specially for lists of authors and journals. If you use one of these the task is easier than with cards.

A full reference for a journal article consists of:

- The names and initials of all authors.
- Year of publication.
- Full title of article.
- Journal, written in its unabbreviated form.
- Volume number.
- First and last page numbers of the article.

If the references have been written out in full like this, you can easily make the necessary adjustments, usually deletions, to suit the target journal. But if you have written abbreviated references it can be infuriating to be obliged to seek the original references a second time for something as trivial as, say, the final page number.

Often reference is made to books as well as journal articles; the complete format for a book reference is:

- The names and initials of all authors.
- Year of publication.
- Title of book.
- Names and initials of editors.
- Number of the edition if there is more than one.
- First and last page numbers of the chapter being referenced.
- Name of publisher.
- Town of publication.

Again, journals vary in their specific requirements but these can be accommodated by appropriate deletions from the list above.

Other forms of reference are sometimes needed. If you are referring to a thesis, and the work you are citing has not been published elsewhere, your reference will have the following format:

- Name and initials of author.
- Year of submission of the thesis.
- Title of the thesis (MSc or PhD thesis).
- Name of university.

University theses are very difficult to consult and if you know that the author has published material from a thesis in a scientific journal you should refer to that rather than to the thesis. If you are not sure, write and ask.

If your attention has been drawn to a piece of research through another paper or book or through an abstracting journal, and you cannot locate the original source, very often a thesis or a foreign journal, it should be referenced in the accepted way but the words 'Cited by . . . ' or 'Abstract in . . . ' should be added to the reference. This should be done only as a last resort when all attempts to find the original source have failed. After all, you are now asking the dedicated reader, to whom the original article may be important, to do what you have failed to do.

Unpublished work or personal communication should be referred to in the text only, and not in the list of references. Editors, particularly, seem to dislike this form of citation in the text so use it only when there is no alternative. Certainly, authors who refer to their own work with the words 'pers. comm.' are provoking the editor unnecessarily.

The Summary

The Summary, which may be placed at the beginning or the end of the paper, depending on the requirements of the journal, is the first thing a reader will normally look at after being attracted by the title. It is therefore much more than an afterthought even though it is possibly the last section you actually write. Whether or not readers will read your paper in full will depend on their impressions after reading the Summary. It should thus:

1 Expand the title.
2 Condense the paper.

There are two sorts of Summary: the informative and the descriptive. The Summary of a scientific paper must always provide information. It must be factual. Statements like 'The results are discussed in relation to published work' are completely worthless.

The four factual elements of a good Summary are:

1 Why you did the experiment.
2 How you did the experiment.
3 Your main results.
4 Your principal conclusions from the results.

When it comes to expanding on these four elements you are in luck, because you have already done the hard work to get the information for each of them so you should to be able to complete your summary very quickly.

Why you did the work can be conveyed in no more effective way than to restate the hypothesis or hypotheses that you tested.

How you did the experiment can be covered by paraphrasing the 'experimental procedures' section of your Materials and Methods.

Your main results are those to which you gave highest priority when you were sorting out the order and emphasis of your Results section.

Your principal conclusions are, in like manner, those that scored highest when you were assessing priorities for your discussion points.

Ideally, each element should be treated in one, and rarely more than two, sentences. Readers want a quick, broad outline of what happened. If they want to know the reagents that you used, the number replications, or other details, they will read the full paper. If your most important result does not interest them, then your less significant results are of no consequence at all. If it does interest them, they will be encouraged to seek fuller details in the main text.

Descriptive Summaries are reserved for review articles and those that synthesise already published information such as teaching or extension articles. They consist of a brief compilation of the main conclusions that you have made progressively throughout the text.

Helpful Rules about Summaries

The Summary is always published with the title—even if the summary finds its way verbatim into an abstracting journal. You can capitalise on this by not wasting space in repeating information already in the title. In fact, in writing the Summary take the title as being the first sentence and proceed from there.

By contrast with the title, the rest of the text of your article may not always be next to the Summary and in any case will not be read at the same time. Therefore, the Summary must be self-supporting. Important information involving differences, means or totals should be given numerically and exactly. You cannot refer to the text or to tables and figures. For the same reason abbreviations should not be used unless the same expression is to be used several times in the Summary itself. The work summarised should be all original and therefore references are not used in summaries.

A short Summary can often be taken verbatim and used by an abstracting service. If it is too long the abstracting service may have to shorten it and without the detailed knowledge that you have, may make a mess of it. So the Summary is not only to induce readers to spend their time with your main paper, but it also assists in the widespread and accurate dissemination of your paper through the abstracting services.

Notes

[1] Beveridge, W.I.B., The Art of Scientific Investigation, Vintage Books, New York.
[2] Ehrenberg, A.S.C., 1982, A Primer in Data Reduction. An Introductory Statistics Textbook, Wiley & Sons, Chichester.
[3] Sokal Robert R. and Rohlf, F. James (1981), Biometry, W.H. Freeman, San Francisco.

Second Draft—Getting it Together

The 'fluency' test

YOU NOW HAVE the basis of your paper in a number of separate sections. They probably took you several weeks or even months to prepare unless you are particularly fluent or had a lot of time to devote to writing. It is likely and quite normal that some sections do not match with others as well as they might. If possible, leave the whole thing for a few days so that when you take it up again it will seem more like a fresh piece of writing. If it has been typed, so much the better, because it will appear to be even less familiar. At this point all that you wished to say will have been safely committed to words and, for the first time, you can now attempt to comprehend the whole structure of your paper in one pass. To do this you must read it as quickly as possible from beginning to end. This is the best test of fluency.

When you come to parts that grate, or that duplicate others, or appear wrong, do not stop to correct them. Put a pencil mark in the margin and pass on because, at this point, your aim is to see how the paper flows. Later you can come back to the parts that did not read well. During the 'fluency test' you will have identified sentences that are convoluted or are too long. If they appear awkward to you, the author, there is little hope for them when they are read by strangers. You may therefore have to reconstruct some individual sentences or alter the sequence of groups of sentences. However, this time you have a great advantage over the last time you wrote them in that you know exactly where the individual sentence fits in to the whole article and its overall contribution. Improving them is much easier on a one-by-one basis.

Duplication

Now is also the opportune time to discover and eliminate unnecessary duplication. The danger points for repetition are usually found in the Introduction and the Discussion, and the Results and the Discussion.

Points developed in the Introduction are often redeveloped needlessly in the Discussion. Where this happens, it is usually possible to amend the Discussion so that the reader is reminded of the argument in the Introduction but is not obliged to re-read it. Results are very frequently repeated in the Discussion. As we saw earlier, the remedy is to refer in the Discussion to appropriate tables and figures, not to repeat text. If this cannot be done you should re-examine the way in which the results have been composed and reorganise the presentation so that you can refer to the sections in question easily and concisely.

Having organised the overall layout of the paper so that it flows logically, we can now begin to examine ways of improving the writing itself.

3

Third Draft—Readability

ENGLISH is the international scientific language. It is frightening to realise that many foreign scientists' only contact with our language is through reading English language scientific journals. In writing English, therefore, we have two responsibilities: to communicate our research findings, and to set an example for fellow scientists, foreign and native-English-speaking. The key to meeting both responsibilities is to construct readable sentences and, ultimately, clear paragraph units. Readable sentences are the elements that generate fluency and coherence in writing. They are sentences that can be read in one pass and understood by someone of average intelligence and with no special knowledge of the subject matter. 'In one pass'; that is the key to readability. The English language is versatile enough to allow us generally two or more options for expressing the same thought. The decision as to which to use should rest on readability. The sentence that can be read and absorbed without having to backtrack or to stop and select between alternatives is the one to choose. Even sentences that are grammatically perfect and unambiguous are sometimes constructed in a way that requires several attempts by the reader to find their exact meaning. Such sentences should be reconstructed in a way that no longer distracts the reader. After all, the words you use are merely the vehicle to carry your thoughts. If the words are unobtrusive but, at the same time, convey the thought accurately the reader has the privilege of only having to concentrate on one thing. Different readers often seek different information from an article and therefore their trains of thought will not be identical. So, it is impossible to meet the criterion of perfect readability 100 per cent of the time. But, at the other end of the scale, an article that requires reading and re-reading, sentence after sentence, becomes a total failure because the reader gives up.

Strangely, when readers find themselves in the midst of a particularly difficult piece of text their first reaction is to think that their inability to seize its meaning is their own fault. 'I can't seem to concentrate to-day'; 'I've too many other things on my mind'. Often the cold truth is that the author is the sole culprit.

Many writers in attempting to meet the constraints of scientific truth and exactitude become tense and stifle their writing style and terminology to produce a heavy, ponderous result. They abandon a conversational style, presumably to convey an impression of the seriousness of science. They may tell you that they think Smith's results are wrong but will sit down and write 'The present authors believe that their results and those of Smith are at variance', because they think that sounds more scientific. If you are writing and find yourself in agony trying to sort out the wording for one or two sentences your problem can often be resolved in seconds when a colleague unexpectedly asks 'Tell me, what do you really want to say?' The conversational reply to that question can supply the answer that has been so elusive; a simple direct statement of what you wanted to say. The momentary drop in tension starts the breakthrough. If we can overcome, as a matter of course, the magnetic attraction of 'scientificese', one of the great barriers to clear writing disappears.

Scientists are not alone, of course, in their attempts to develop a style of writing that they think befits their calling. Lawyers, clerks, journalists and others have all contributed to the convolution of the English language. If you wrote to your local council and said:

> I can't see the traffic when I back out of my garage because of a tree in the street in front of my house. Would you please remove it?

Do you think they would understand you? Many people apparently think not, because they try to help out the poor people on the council by constructing something more 'councilish'.

> The writer wishes to make a request in respect of an obstruction on the border of the thoroughfare in front of the writer's residence. The obstruction—viz. a large tree—impairs vision of vehicles on the said thoroughfare in relation to entry and exit from the writer's residence. I

hereby request that the council take steps to immediately remove the said object at the first available opportunity.

Surely the council staff are ordinary people capable of understanding simple English.

Scientists, too, are ordinary people and neither need nor appreciate contorted English.

There are, therefore, two sources of errors of style in scientific writing. The first stems from the writer's lack of understanding of English syntax either because English may be a second language or because grammar was a weak subject at school. This is a deep-seated problem for which the reasons are obvious and cannot quickly be overcome.

The second stems from the tension of simply trying to be scientific. This tension expresses itself in two ways. The writer inserts cumbersome expressions into the individual sentences of the text in attempts to make each sentence a scientific mini-masterpiece. Then, having expended so much effort on individual sentences, he or she fails to recognise that those sentences must relate logically and fluently to each other.

Cumbersome expressions

Our concepts of writing style are learned largely from what we read in scientific journals. Unfortunately, what we read is not always the best example on which to base what we write.

Here are ten common examples of convolution of English that demonstrate a sort of scientific style but which do not result in a clear communication of facts. In each case, the offending construction is bad because it interferes with readability. The reader has to stop, untangle what is written, decide what is meant, and only then, read on. Look up your most frequently used scientific journal and see how long it takes you to find two examples of each of the ten problems illustrated below. It won't take long.

1 Clusters of nouns

Noun-noun-noun sequences are probably the most common form of scientific jargon: for example:

Random leaf copper analyses

Difficult child psychology problems
Amino acid digestion analyses.
Chemical healing suppression figures
Third generation portfolio planning

′ These expressions pose two problems. They are always cumbersome to read and they are often imprecise. Sometimes they are used in the belief that valuable space is saved by eliminating prepositions such as 'of', 'on', 'in', 'for' and others, and sometimes writers have become so used to thinking of a certain group of nouns as one, that they do not realise that a reader trying to assimilate them for the first time will be floundering. Occasionally, as in the case of 'third generation portfolio planning', they are deliberately designed to confuse with words that imply that the writer has some privileged knowledge that is not available to the reader.

Omitting prepositions may be permissible where the missing word is clearly understood. But, does 'chemical healing suppression' mean 'suppression of healing *by* chemicals' or 'suppression *of* chemical healing' by something else?

Where several nouns are clustered and there is also a real adjective in the cluster the reader can often confuse the noun to which the adjective refers. To illustrate, in the first example, are we dealing with random leaves or random analyses?; or in the second, difficult children or difficult problems?

There are three possibilities here.

a Replace one or more of the nouns used as an adjective by the real adjective; for example, 'psychological problems', not 'psychology problems'.

b Use the appropriate prepositions; for example, 'random analysis of copper in leaves', or, of course, 'analysis of copper in random leaves'. By a happy coincidence, prepositions are among the shortest words in the English language. Inserting an extra one or two to be more accurate will not lengthen your article significantly.

c Where words seem particularly appropriate together—and this is comparatively rare—use a hyphen to indicate that they should be read as one composite noun, for example, 'healing-suppression'.

2 Adjectival clusters or adjectival clauses

For example:

The analysis was carried out to find *the maximum net returns above feed cost ration.*

Research in the manufacturing industries has operated on a 'grant' culture rather than *an innovation based return on investment culture.*

These produce the same problems in readability as clusters of nouns. They are generally the result of an over-familiarity with the field and generally confuse readers who see them for the first time.

3 Sentences beginning with subordinate clauses

These sentences are the hallmark of pseudo-scientific writing. They seem to be used to indicate that the writer has taken care to clarify the main clause by first declaring all reservations about it, thereby implying precision. The impact is entirely lost because no one knows what is being clarified until they get to the main clause which may be several lines further on. Our short-term memory just can't handle it. The most important part of any sentence is the beginning and that is where the most important message should be placed; for example:

> Thus, although there were too few plots to show all of the interactions which we sought [subordinate clause, apologetic], under the conditions of the experiment [subordinate phrase conditional], copper and zinc acted additively.

Compare that with:

> Thus, copper and zinc acted additively under the conditions of our experiment although there were etc . . . ,

which is much easier to follow because we know what the sentence is about (or the topic) from the first few words.

Other examples are sentences beginning with: 'Despite the fact that . . . '; 'Notwithstanding the fact that . . . '; 'While . . . '; 'Whilst . . . ' (meaning although). These words signal the beginning of a sentence that is likely to be difficult to comprehend in one pass.

Occasionally, a condition or reservation may be the key issue in a sentence. In this case you are justified in placing the conditional clause first. For example, after a statement about the value of fertilisers you may say: 'If there is insufficient rainfall it is uneconomical to apply supplementary fertilisers'. Nevertheless, on the rare occasions when you put a subordinate clause first, be sure you do so for the right reason.

4 Nouns instead of verbs from which they are derived

For example:

'Weights [noun] of the animals were taken';

'Low temperatures caused a reduction [noun] in the rate of the reaction';

'Recording [noun] of pulse rates was made';

'Temperatures showed an increase [noun] during the day'.

One of the most successful methods of repairing such sentences, all of which seem cumbersome, is to look at each noun in the sentence and see if it has a verb derivative. If so, simply use the verb. Thus:

'The animals were weighed' [verb] or 'we weighed' [verb] the animals'; or

'Low temperatures reduced [verb] the rate of the reaction'.

You will notice that by creating a new verb from a noun we have done three useful things. We have automatically dispensed with the original verb, which indicates that it didn't have much value in the first place. We have shortened the sentence and we have sharpened its impact. Replacing nouns with verbs is one of the most simple and yet most powerful ways to improve the clarity and directness of your writing.

5 Use of 'filler' verbs

Verbs in this category are often added to complete a sentence in which the appropriate verb has been wasted because it is in its noun form. For example, in the statement:

'We conducted a study of pathogenic insects.'

we can use the verb 'to study' ('We studied . . . ') and the verb 'to conduct' disappears. 'To conduct' was a 'filler' verb.

'An improvement in digestibility occurred when an increase in the protein content of the diet was made.'

This becomes:

'Digestibility improved when the protein content of the diet was increased.'

The sentence is shorter and clearer in the absence of the two filler verbs 'to occur' and 'to make'.

There are many verbs in this category, of which some of the most common are 'to occur . . . '; 'to be present . . . '; 'to be noticed . . . '; 'to obtain . . . '; 'to take . . . '; 'to perform . . . '. They are so non-specific that you can often substitute one for another and make no real difference to the meaning of the sentence. Whenever you see them in a sentence, look for a noun in the sentence whose verb derivative you might use instead. The modified sentence will invariably be clearer and you will notice that there will be no other verb that you can satisfactorily substitute for the new word that you used.

6 Use of passive rather than active voice

Passive voice is useful when the doer of an action is not known or when it doesn't matter who or what performed the action. In all other cases it makes the expression wordy and vague when compared with the more direct and straightforward active voice. When you use passive voice in a scientific article to describe your methodology, or to express an opinion, one could cynically suggest that it implies that you do not want to be held responsible for doing the work or having that opinion. For example: 'Patients were observed [passive voice] by two people for signs of abnormal behaviour . . . '; 'It is believed [passive voice] that, in this case, chemical analysis is better than bioassay'. Changing to the active voice they read: 'Two people observed the patients . . . ' and, 'I believe that . . . '

Many people, when using the active voice, see the use of the first person—'I' or 'we'—as a problem . They think that it destroys objectivity and that more distant words like 'the author' are somehow preferable. So, we get sentences like 'The author disagrees with Bloggs (1989)'. One or two journals frown on the use of the first person but most do not. I believe that the use of the first person and active voice gives a refreshing sense of directness and involvement and sometimes avoids the necessity for some remarkable verbal gymnastics. 'He was told by the author that the lake should be jumped into by him.'!

If it is immaterial to the sense of the sentence whether it was you or anyone else who performed the action then, by all means, use the passive voice.

7 Use of imprecise words

These are words like, 'considerable', 'quite', 'the vast majority', 'a great deal', 'rather', 'somewhat', 'etc.' and 'and so forth'. Each of these words can convey a different meaning to different readers. Considerable could mean anything from a few per cent to ninety-nine per cent. It is invariably more specific and more useful to give the exact figure or a rounded version of it. Thus, instead of 'A considerable number of plants responded' we should use 'Seventy-four per cent of plants responded', or even 'About three-quarters of the plants responded'. The rounded version should be used only if the precise figure is given in an accompanying table or figure.

Words like 'etc.' and 'and so forth' are often used when the writer cannot think of anything more to complete a sequence of words. This is the very antithesis of scientific precision. 'The data were treated statistically to take account of changes in temperature, humidity, daylength, etc.' Can you guess what 'etc.' means here? Avoid 'etc.' as a matter of course. If you insist, then the only time that you should use it is when the identity of the 'etc.' is absolutely clear. 'The 20 aliquots were labelled 1,2,3,4, etc.'

8 Use of compound prepositions

For example, 'in the case of', 'in regard to', 'as to whether', 'in respect of'. These are simply padding and dilute the meaningful parts of the sentence. They are the stock-in-trade of speech makers, university lecturers and politicians who use them to gain valuable seconds while thinking of the next thing to say.

9 Use of multiple negatives

For example, 'it is not uncommon', 'it is unlikely, that it won't work', 'not unreasonably inefficient'. Two negatives make a positive in both English and mathematics. Why not save the reader the trouble of calculating and be positive in the first place?: 'it is common', 'it is likely to work'. I defy anyone to be certain that they have the right meaning for 'not unreasonably inefficient' without hesitating and recalculating several times. As someone once said, 'People who use double negatives make me not unill!'

10 Use of unfamiliar abbreviations, symbols and references

Included in this category are all those things over which readers are likely to stumble, and which might be expected to break their train of thought.

Abbreviations

Abbreviations frequently require a moment or more of consideration even if they have been explained earlier in the article. Abbreviations can certainly be useful—especially if expressions that could be abbreviated are to be used many times in a paper. Even so, they should be written out in full in the title and in headings to graphs or tables. In short, anywhere that they might be encountered separately from the text in which they are defined. This also allows the reader more opportunity to assimilate them.

But don't go wild with abbreviations because they can be particularly annoying and distracting. If an expression is not used more than three or four times, the saving in space through abbreviation will in no way compensate for the readers' lost time and concentration while they verify the meaning of the abbreviation. Commonly accepted and well-known abbreviations—which may not be as commonly accepted or as well-known as you imagine—are usually difficult enough for most readers. For example, AA means 'amino acid' to biochemists and 'atomic absorption' to physicists but it is also familiar as 'Automobile Association' to motorists, and 'Alcoholics Anonymous' to others (presumably not motorists!). Abbreviations that you invent yourself should be avoided except as a last resort because they invariably disrupt readability.

If you are an endocrinologist you might understand this: 'FSH and LH were measured by RIA and E2, was extracted with RTC, purified by TLC, and measured by CPB'. If not, you would need several minutes at least to begin to comprehend what was being said. There has been a trend in recent years for government documents and consultants' reports to include a large table of abbreviations at the beginning for the benefit of the reader. I cannot think of a more obvious way of signalling that the document is going to be ponderous to read. In effect, readers are being warned that they will be obliged to stop reading each time that they encounter one of these monstrosities, refer to the table at the front for clarification and then try to pick up the thread of the article again in the body of the text.

On the same criterion of readability, expressions like 'kg day-l' instead of 'kg per day' seem unhelpful. We say 'kilograms per day', not 'kilograms day to the power-1'. Therefore, it seems preferable to write what we say no matter how mathematically correct the other expression. A venerable colleague of mine when he first saw that cows were being fed a ration at the rate of 10 kg day-l suggested that they were probably being fed at night.

Referencing

We have seen already that the positioning of references to published work in the text can convey subtle differences in emphasis, but references should not be allowed to break the flow of sentences unless for a special reason. Consider this sentence:

> The number of stomates per leaf may increase in geraniums (Brown 1937), decrease in petunias (Black 1978) or remain constant in sweet peas (White 1990) when manganese is deficient.

The construction makes sure that each fact is accorded its appropriate author but the sentence is difficult to read because the authors have intervened unnecessarily. A more acceptable statement, because it is more fluent, is:

> When manganese is deficient the number of stomates per leaf may increase in geraniums, decrease in petunias, or remain constant in sweet peas (Brown 1937; Black 1978; White 1980).

Note also that the key to the sentence—'When manganese is deficient'—has been placed at the beginning even though it is a subordinate clause. This way the reader immediately grasps the perspective of the writer.

Footnotes

Very few scientific journals encourage, and fewer still allow, footnotes with explanatory statements or references at the bottom of each page. In my opinion, that is a very good thing. Even where such major distractions are permitted, you should avoid them in the interest of your reader. Certainly, they provide extra information at a particular point in the text but at what cost and to what end? The reader is lured from the mainstream of the thought process to a side issue, with potentially disastrous consequences for his or her comprehension of the major text. But, was the side issue worth pursuing anyway? The fact that you, the writer, considered it unworthy of a place in the main body of the text suggests immediately that you, the writer, should consider very seriously leaving it out entirely. If it is important enough, then put it in the main part of the article; if it isn't, then leave it out. Messing around with footnotes on the assumption that some people may be glad of the extra information is effectively admitting that you don't know why people are likely to read what you have written. That, in turn, means that you don't know why you are writing the article in the first place.

Organising sentences so that they are readable

Helpful rules in organising your writing

1 Power of position

The English language is remarkably flexible. Most sentences contain a series of pieces of information or ideas and there are many ways in which these can be expressed. The choice of the best of the many ways depends on which of the facts or ideas is most important to the development of your story.

Fleming, in 1929, discovered penicillin after a bacterial plate he was culturing became contaminated with a spore of the fungus *Penicillium*.

Some of the many facts and ideas in this sentence are:

1 The discoverer of penicillin.
2 The date of discovery.
3 The way it came to notice.
4 The name of the organism involved.
5 What it contaminated.

It is likely that one or other of these pieces of information would be more important than the rest depending on your objective in writing the sentence. For example, if you wished to emphasise that it was Fleming who made the discovery, you would probably be happy with the sentence as it stands. However, if you were emphasising the historic implications, the date would be your main consideration. The sentence would be slightly modified to 'In 1929 Fleming discovered penicillin . . . ' If you were in the process of describing the various antibiotics, you would want the drug penicillin to be emphasised and so your sentence would read 'Penicillin was discovered in 1929 by Fleming after . . . ' If you were describing the role of accidental discoveries in science you would construct your sentence in another way: 'After a spore of the fungus *Penicillium* contaminated a bacterial plate he was culturing, Fleming . . . ' You will have noticed that, in each case, the desired emphasis has been conveyed by placing the major element first in the sentence.

The most powerful position in any sentence is the beginning

The beginning of the sentence should be held sacrosanct and reserved for orientating your reader. The beginning of the sentence gives readers their bearings and enables them later to pick up new concepts or the less essential details that follow, easily and without any irritation. It is the 'topic' part of the sentence. Bearing in mind that your aim is to keep the reader as close as possible to your thought path, a knowledge of this simple rule is one of the most powerful tools you can have in your repertoire. We have already seen that the concept of the beginning being the most powerful position applies equally to sentences, titles, paragraphs and to whole sections of the scientific article.

A scientific article that presents all of the data and all of the scientific discourse that the author intended to present is not necessarily a successful article. It only becomes one when most of the people who read it can perceive accurately what the author really meant. For this to happen the author has to be aware of what makes things easy to read. As Gopen and Swan[1] (1990) observed, 'If the reader is to grasp what the writer means, the writer must understand what the reader needs.' Gopen and Swan describe a methodology to achieve this based on the concept of 'reader expectations'.

Basically, all information that we receive by the written word is either 'new' or 'old'. That is, it provides us with fresh concepts and ideas or else it consolidates ideas that we have already received. In most cases, we can find both of types of information in the same sentence. The key to rapid comprehension is to use the 'old' information to let readers know where they are in relation to what they have just been reading, and then present the 'new' information.

New thoughts are grasped much more readily when they are perceived from the comfort of what is already understood. So, the first part of the sentence should usually be used to get readers comfortable by linking them to previous information before the rest of the sentence discloses the new idea. The order is most important and we can often make great changes in the readability and the clarity of passages simply by getting the order right. If, at the same time, we take care to provide linking words that signal to what our next idea is going to relate we can almost work miracles with seemingly difficult text. The value of this idea of generating in the reader an expectation against which he or she can compare new information cannot be stressed enough. The concept works at the level of the whole article, the section, the paragraph or, in this case, the sentence.

Consider the following paragraph which describes how fat is broken down in our small intestine. It is taken from a published textbook on Physiology.

The main fat that we eat is the ester of long chain fatty acids and glycerol. The small intestine is the site of digestion and absorption of fat. Under normal conditions we do not excrete fat in the faeces because almost all of the fat that we eat is absorbed. Fat leaves the stomach as large droplets within an aqueous solution of chyme. If it remained in this form the water soluble lipase which digests it would have great difficulty in getting into contact with most of it. Bile salts can emulsify large droplets and break them down into smaller ones and the lipase can come in contact over a much larger area. Agitation within the duodenum helps break up the large droplets, the lipid part of the bile salt molecule dissolves in the fat and the electric charge on the polar part of its molecule faces outwards towards the aqueous phase of the mixture preventing the droplets from coalescing.

Most people would describe this paragraph as heavy going. Yet none of the sentences within it is particularly difficult to read. They are grammatically correct, they don't contain large and obscure words except those that are appropriate scientifically and, apart from the last one, are not inordinately long. The problem is that when taken together they make us work too hard to follow their overall sense. Each sentence assaults us with new material without regard to what the sentences around have been telling us. We haven't had an opportunity to get 'comfortable' with old material before new material is thrown at us. Our minds have difficulty in pigeon-holing the information in a logical way and this leads to at least two unsatisfactory consequences. The first is that readers are obliged to store a lot of information on 'hold' while they backtrack and re-read to find more clues about what to do with it. The second, as a direct result of this confusion, is that the material is likely to be interpreted by different readers in different ways.

What are the new bits of information in each sentence of this paragraph?

The main fat . . .

We don't normally excrete fat . . .

It is in large droplets . . .

Lipase can't get to it . . .

Bile salts emulsify large droplets . . .

Agitation helps . . . so do electrical charges

In most cases these new pieces of information are being made at the very point in the sentence where the reader is unprepared to receive and absorb them—at the beginning. The sentence needs to be made more 'user friendly' to allow the reader's mind to tidy up and put away the material from one sentence and prepare for the next. Here is my attempt to apply these concepts:

> The main fat that we eat is the ester of long chain fatty acids and glycerol. We digest this fat in the small intestine and absorb it almost totally because we normally do not excrete fat in the faeces. When fat enters the small intestine *from the stomach* it is in the form of large droplets within an aqueous solution of chyme. The aqueous layer acts as a barrier to the water soluble *enzyme*, lipase, preventing it from contacting most of the fat and digesting it. *So the barrier must be broken down and this is done in three ways. First*, large droplets are emulsified and broken down into smaller ones by the salts in bile *which is excreted in the small intestine*. The lipase can now come in contact with fat over a much larger area. *Second*, agitation within the small intestine helps break up the large droplets. *Third, the bile salt molecule has a lipid part and a polar part that is electrically charged.* The lipid part dissolves in the fat and the electrically charged part of its molecule faces outwards towards the aqueous phase of the mixture preventing the droplets from coalescing.

This is easier to read because it now has a structure that presents new information only when the reader has been made ready to accept it.

The first sentence has not been changed because, in the absence of a preceding paragraph we have no 'old' information on which to build. It serves as a topic sentence for the new paragraph. But, in the second sentence the new information about the small intestine is not raised until we have linked it with the old information from the first sentence. The modified sentence now begins by establishing that it is going to continue to tell us about fat. Similarly, the third sentence now orientates us towards the now familiar small intestine before introducing new material about fat droplets. A logical flow from one sentence to the next has been built up and continues throughout the paragraph.

The modified paragraph is longer than the original because some new material, highlighted in italics, has been deliberately introduced. This is the direct result of considering the sequence of events as the reader might perceive them. In the original, the author had neglected to tell us about some of the logical connections so they have to be assumed by the reader. In the new version, these omissions have become obvious and have had to be inserted. The author was probably so familiar with the fact that food enters the small intestine from the stomach, that lipase is an enzyme, and that bile is actually excreted into the small intestine that it seemed unimportant to say so. Certainly, some readers may agree, but many others would not. Unless the author can be sure that all readers are as informed as he or she is, then it is a wise assumption that they might need a little help.

A further invaluable aid to developing the expectations of the reader is illustrated in the last part of the new paragraph. It has been made clear that there are three ways of breaking down fat droplets. The simple, short sentence saying so is a map that keeps the reader orientated through a relatively complex passage of information and keeps each piece of that information in perspective.

2 Checking the tense

The rules for the choice of tense of verbs are relatively simple. In almost all scientific articles only two tenses are used, the past, most of the time, and the present, sometimes. The only exception is when a sequence of events in time is to be described and each event has to be placed relative to the others. Then the pluperfect or the past continuous tenses might be used. For example:

> After the patients had had [pluperfect tense] a barium meal they returned [past tense] to the operating theatre.

> While the plants were wilting [past continuous] they lost [past tense] their nutritive value for livestock.

Otherwise all descriptions of what you did and your results are described in the past tense. The reason is that your experiment is now finished.

The present tense is reserved for two conditions. Conclusions, generalisations and principles that you believe are still valid at the time of writing, and 'housekeeping' within your article where, for example, you refer to tables or figures. 'Figure 3 shows [present tense] that . . .'

When describing the work of others the same rule applies. For example:

MacSpratt found [past tense] that when sheep were [past tense] deficient in nitrogen the rate of mitosis in wool follicles was [past tense] sixty-three per cent of normal and concluded [past tense] that nitrogen is [present tense] essential to normal growth of wool. This is [present tense] shown graphically in Figure 4.

Each of the verbs in the past tense describes a specific event or result; the first verb in the present tense is part of a generalisation and the second involves a piece of housekeeping. In this example, the verb 'is' in the generalisation could be replaced by 'was' if, in fact, later work had proved the conclusion to be false.

Let us look at another example:

Infestations with aphids reduced [past tense] the yield of raspberries by 18 per cent and treatment of the aphids with pyrethrum was [past tense] 98 per cent effective. Pyrethrum is [present tense] cheap so that it can be used [present tense] to increase yields of raspberries economically.

Again the distinction between the descriptions of events, which take the past tense, and principles, which use the present tense, is clear.

3 Precision, clarity, brevity

These three criteria, more than any others, distinguish scientific writing from other forms of literature. Not only must they constantly influence how a scientist writes, but they must always be considered in that order. It is good to be brief but if, in so doing, you do not express yourself clearly then brevity should be sacrificed to achieve clarity. Similarly, precision should never be sacrificed in order to make it easy to say something clearly or more briefly.

The colleague test

Despite your efforts, it is unlikely that you have been entirely objective in your appraisal of your own work. It is equally unlikely that expressions and explanations that seem adequate to you will be equally clear to someone else. Now is the time to seek help from some sympathetic and respected colleague. Ideally, find at least two people who will read your manuscript and make frank comments about it.

Ask them to read the paper quickly, as you did, marking difficult passages in passing without attempting to fix them. A reader who is familiar with the field of work or who may even have had a small part in the experiment should be able to make constructive comments about the substance of the paper and the correctness of your arguments. If possible find a second reader who is familiar with scientific literature but not with your field of work. His or her comments on the fluency of the paper, its comprehensibility and the presence of jargon and awkward abbreviations should be taken very seriously. If they say that they do not understand a section you should take the view that this is your fault and not theirs. It is not enough to show them where they misinterpreted you or failed to grasp the meaning of something. The fact that they did not grasp your meaning at the first attempt probably means that it was not expressed as well as it could have been. Therefore, you should try to reconstruct the offending passage to prevent other readers from having the same difficulty.

Note

[1] Gopen, George D., and Swan, Judith A., (1990) The Science of Scientific Writing, American Scientist 78, 550-558.

4

Final Draft—Editing

Meeting requirements of the journal

E ACH JOURNAL has its own 'house style'. That is, it requires abbreviations of commonly used units to be uniform, headings to be set out in a particular way, references to be cited in the style of the journal and so on. This information is given periodically in the journal itself or in the case of some journals in a separate 'Guide to Authors'. Some of the points appear trivial but editors are looking for uniformity in their journals and insist that their format be followed carefully. To be certain that you take nothing for granted it is a good idea to make a photocopy of the journal's instructions to authors and keep it beside you at this stage. In addition, as you, or a typist, prepare the final draft a copy of the journal itself on the desk can often solve minor problems as they arise.

Checking

The next job is verification. Spelling-checkers in word processing programmes can help you pick up some misspelt words and typographical gaffes. There are programmes that can check your grammar and syntax. These are certainly helpful at this stage of the writing process, but they do not and cannot substitute for meticulous checking of both the text and the figures.

Your scientific credibility depends on many things but above all on your exactitude. There are many sources of error in science associated with variability, chance, and limitations of available techniques. Statistical techniques have been developed to preserve your credibility in the face of these but if you introduce another source of error—plain carelessness—you have little recourse to aid—or sympathy. It has now been a long time since you first began manipulating your data from their original form. The chances of an error of calculation, transcription, or typing are therefore high. As well as checking the final draft against the original data you should also recalculate the means and totals that were derived from them. Then check the text of the Results and Discussion sections to be certain that the figures quoted correspond to those in the tables or graphs. It is equally important to make sure that each figure or table is referred to in the text by its correct number. To stress the importance of verifying such obvious things hardly seems necessary. But the time spent is well worthwhile if it enables you to avoid the embarrassment of having a manuscript returned because of a glaring clash of figures or with an unfavourable report based on mistakes in your data. A further consequence is that reviewers who find careless errors, no matter how obvious, are inclined to suspect other more subtle mistakes and submit an appropriately guarded, if not damning, report on your work. Thus, verification is not a step that involves ten minutes between other jobs. It must be done slowly, meticulously, and fully, with pencil in hand and mind alert. Here is a check list that should be followed with the same precision as a countdown for a rocket launching.

Checklist for the editing stage

Step 1 Has your draft paper been read and criticised by:

1 a colleague from your field?
2 a colleague from another field?
3 a person fluent in English (especially for non-native English speakers)?

Step 2 Have you selected the journal in which you hope to publish this paper and made a photocopy of the instructions to authors?

Step 3 Using the journal's guidelines for format (step 2), have you checked:

1 your Reference section?
2 the references in the body of the paper?

Step 4 Have you revised, taking note of steps 1 and 2:

1 your Title?
2 your Summary?
3 the format of your headings, sub-headings, key words and running headlines?
4 all the points criticised in Step 1?

Step 5 Have you now rechecked:

1 the precision of your references, by looking again at your original sources either in the library or in your folder of photocopied materials?
2 the agreement between sources referred to in the paper and those listed in the References section?

Step 6 Have you proofread the final typed paper and checked it for:

1 omissions from your original text?
2 typing errors, especially spelling, numbers, formulae, tables and graphs?

Problems of publishing

After making certain that you have taken care of all the little things that the chosen journal insists on—like numbering lines, referencing and headings—you are now ready to send your work to the editor. All you need is a simple covering letter asking him or her to consider your paper for publication and telling them where to send correspondence if you have changed address or if there are several authors with different addresses.

What happens when the editor receives your manuscript? The editors of most scientific journals these days are employed full-time to maintain smooth production of the journals and safeguard their standards. The editor's first step is to check the subject matter of your paper to verify that it falls within the normal scope of the journal. Then he or she selects one or more, usually two, referees who work in a similar field to yours and who can give guidance on the scientific merit of the paper. Your paper is sent to these people with instructions that they comment on the soundness, the originality, and the relevance of the work. Editors normally do not seek commentary on the style of writing or the layout because that is their responsibility as editors. After the editor receives the referees' comments he or she reads the paper carefully to ensure that its style is compatible with that of the journal. They then write a letter to you with one of three verdicts: acceptable to be published; acceptable with certain minor or major modifications, or unacceptable. In each case the editor's own comments on the manuscript are sent to you together with those of the referees to justify the decision. In most cases the identities of the referees are not given.

The arrival of the editor's letter represents for you the moment of truth after a great deal of work. Depending on which of the three decisions the letter contains you have three courses of action.

The paper is accepted

You can feel very proud of your work because few papers are accepted without some changes. The perfect paper has probably never been written so there are always at least a few improvements that can be suggested after scrutiny by three experts.

The paper is accepted but with modifications

After spending long hours over a manuscript writing, modifying, checking and re-writing, it becomes almost part of you. It becomes a child of your creation. Parents never like someone saying that their child is ugly or imperfect. Therefore the initial reaction to harsh comments on the substance of your paper is almost inevitably one of anger and disappointment. You wish to dismiss one, two, or all three of your critics as fools and take your business elsewhere. That is clearly not the mood in which to construct your reply to the editor. Instead, it is wise to leave the manuscript and comments in a drawer for a few days until your disappointment subsides. In thirty years of going through this process I have never ceased to be amazed and annoyed that in about seven cases out of ten I have had to admit, once I have calmed down, that the referees or editors have been completely right.

In another two out of ten they have been wrong only because I have expressed myself badly, and in only about one case in ten have I felt justified in standing my ground. I have to rationalise my disappointment by reminding myself constantly that of the thousands of thoughts and ideas that go to make up a paper only six or eight have to be proved inadequate for the paper to seem poor. No doubt—I tell myself—if the referees had had to start writing the same manuscript starting with a blank sheet of paper as I did they would have made at least as many, but presumably different, mistakes.

It is wise to fortify yourself in some such manner, repair the nine out of ten faults in your paper and have it re-typed. This time, however, your covering letter to the editor must be much more comprehensive than before. You must explain in detail, point by point, how you handled each criticism made by the referees. For those points that you wish to challenge you must be able to give the editor watertight reasons why the referee's reasoning was at fault. The editor, at this stage, acts as an arbiter between you and the referee. If your argument is sound it will be accepted. There is, after all, no reason for the editor to believe that the referees are any wiser than you are and so it is possible to counter their criticism by sound argument.

The paper is rejected

Here editors have had to be very careful. On the one hand they have to maintain the standard of their journals. On the other, they have to say that work that the writer thought worth publishing is, in fact, not worth publishing. This sort of criticism is never pleasant for either the editor or you.

From the writer's point of view all is not always lost. It is important to read the editor's covering letter very carefully. To support a decision to reject a manuscript the editor is obliged to give well-argued reasons. Most editors are experienced and if the letter clearly states that the editor will not accept the manuscript because of his or her own judgement on the matter, there is little to be done. If, however, the paper is rejected on the basis of unfavourable referees' reports it pays to look further.

Many referees, in contrast to editors, are inexperienced. In addition, the cover of anonymity sometimes makes them more dogmatic and less objective than they might otherwise be. One journal which publishes an annual report consulted 470 different referees in a year in which it accepted 171 papers. Only the most naive student of human nature could be convinced that of those 470 people none would have personal prejudices, biases, or even jealousies which could influence their report and hence the editor's decision. If you do feel you have been the victim of such a report you are justified in challenging it. In doing so, you must remember that you will be starting from an unfavourable position and will have to have an argument for every point raised against you. You cannot simply say that the referee is a fool and leave it at that. The exception is where referees show their own brand of poor science by making general criticisms like 'This paper is poorly written'; 'The discussion makes heavy going'; or 'This paper is not very interesting' without immediately following with precise reasons or examples. In the absence of any support from the editor, they have not put up a case for you to answer.

In forming your arguments to meet criticisms, you should never stray far from your hypothesis. Some referees, who are chosen, after all, because they work in a similar field to yours, are apt to follow their own line of thought or pet theory, and expect you to do the same. If this is likely to ruin the integrity of your paper, you can meet the challenge by explaining why it is not relevant to your hypothesis.

The refereeing system is a good safeguard of quality and is preferable to no system of reference at all. But since it relies on human judgement it cannot be foolproof. A colleague recently sent a manuscript to an overseas journal by sea mail. He realised that there would be a long delay and the next day sent a copy by airmail. He later received two separate letters from the editor, one accepting the paper without modification, and the other rejecting it completely. The editor had mistakenly treated the two copies as different papers and his replies were based on the reports of two different referees.

Part II

Broadening the Scope of Scientific Writing

The Review
The Article for a Seminar
Practical Reports and Reviews by Students
The Scientific Thesis
Science for Non-Scientists

5

The Review

THE REVIEW is a special type of scientific article which is in many ways like an extended version of the Discussion of a research article except that your own results are not being discussed.

To write a review we must first know its purpose. Thirty or forty years ago when the volume of literature on any subject was much smaller than it is now, reviews were often written to give a full account of who was responsible for what in a specific field. This enabled interested readers to have a relatively rapid overview of the work done up to the time of writing. Readers could then use the bibliography to look up selected papers for further study. Nowadays this function of the review has practically disappeared. Modern computer-based systems which search the literature can provide bibliographies that are both complete and extensive.

On the other hand, the well-written, modern review has become more important than ever. This is because it is a source of ideas. Computers cannot reason and develop arguments and the purpose of good reviews is to supply this indispensable ingredient in the scientific literature. Let me repeat that. The purpose of a good review is not to present a catalogue of names, dates and facts, but to present reasoned arguments about the field under review based on as many names, dates and facts as are necessary to support those arguments. This means that it is not obligatory to have very large numbers of references. A modern review is judged solely on the quality of its ideas and opinions.

How, then, is a good review developed? It is not an account of a piece of research or an experiment so it contains no new data. Indeed reviews that present for the first time lots of new information in the form of 'unpublished data' or 'personal communication' can be infuriating. Without the opportunity to assess such data and evaluate the methodology behind them in its correct context the reader may well feel cheated. So the good review uses data from previously published material and develops arguments from these. There is therefore no section for Materials and Methods nor for Results. There is only a very simple introduction which does not culminate in an hypothesis as it does in a research paper, but merely serves to outline the scope of the review. The introduction is still very important because one of the difficulties in writing a review is to choose the limits of its coverage. Readers for their part are also anxious to know what aspects of the topic are to be covered and seeks this information in the Introduction.

The format and the layout from this point vary with the topic and its scope and there are few constraints. There are nonetheless three important components. A review must:

- Present new ideas.
- Review all of the literature relevant to these ideas.
- Be specific.

New ideas

An essential feature of a review is that the reader be led to 'the frontiers of science' in the area covered. The most satisfactory way of doing this is by the now familiar method of developing logical arguments until they end in hypotheses. These hypotheses are the core of the review—the ideas that distinguish it from a catalogue of facts. As in every other case, the hypothesis must be supported by the information and must be testable. In a research article the hypothesis is the keystone and must of course be immediately testable with the available technology. In a review, the word 'testable' can be interpreted more liberally. It is not always necessary that present technology be adequate to test the hypothesis. Good reviews sometimes emphasise areas where technology might be improved in order to provide the tools for the advancement of some branch of science. Of course, you must use some discretion in this interpretation of testability. Ideas that are never likely to be capable of being tested are no more than wild speculation.

It is impossible to present new ideas on every aspect of the material you are covering. A coherent review will therefore contain at least some segments of straightforward, factual material that do not lead directly to hypotheses. This does not prevent some interpretation, and often your opinion, based on your knowledge of the field and the value of the information which is available, can help orientate the reader. If you say 'I think Brown (1980) is right because . . . ' or 'Brown's interpretation seems the most feasible because . . . ' you are not supplying new information or even new ideas but you are adding to the interpretation of existing data and theories. Note, however, the importance of the word 'because' in each introductory clause. The presentation of material in the form of reasoned ideas, reasoned opinions, and reasoned judgements stamps the personality and the scientific skill of the author on the review. Thus the data are not firsthand, having already been published elsewhere for the most part, but the review is nonetheless original because it has built on these data to come up with new points of view.

The literature

As far as practicable all of the literature relevant to the field being covered must be presented. We all know that some research data are more reliable than others and it is usual in a good review for this fact to be brought out. Obviously, in developing hypotheses the only data worth using are those that are reliable. However it is inadmissible simply to ignore unreliable data which ostensibly are relevant but which sometimes appear to refute your hypothesis. These data must be presented and soundly repelled by argument or you must explain why you think they are not relevant to your case. Hypotheses based on data selected without good reason are open to immediate criticism and lose their credibility. You may decide that you cannot find a place for some sound data because you do not find them relevant to the development of your arguments. They should certainly be left out if your arguments are to remain clear. Nonetheless, you may decide it prudent to check your Introduction to be sure that in defining the limits of your review you have made it obvious that the scope would not include such irrelevant data.

Another hazard in conforming to the rule that all relevant data be included is that sometimes there are too many references at key points about the same information. Apart from quoting all the references, which is messy and unnecessary, you have two possibilities. First: cite the first author or authors to have made the point in question. Usually the remaining references will have referred to the original article anyway. Second: yours is probably not the first review in this area. If certain points have already been well reviewed with a sound bibliography, you have the acceptable short cut of referring to that review. In these ways the literature can be covered even though some of it will not be in your own Bibliography.

Being specific

The fact that a review covers a wider subject range than a research article is often an encouragement to be unscientifically expansive. Generalisations based on logical reasoning are, of course, an integral part of the scientific method. But generalisations such as:

> 'Extensive investigations are needed to understand the exact role of hormonal, neural, and sensory experiential factors as they affect the reproductive phenomena in farm animals . . . '[taken from a recent review]

are scientific non-statements which should never be tolerated. Either they are so obvious that they need not be stated, or so vague that they have no real meaning. One of the hopes of all reviewers is that they will stimulate others to further research in the same field. The way to do this is to present stimulating ideas, not to indulge in general exhortations.

Some common difficulties with reviews

Suppose your study of the literature reveals two conflicting views on a topic and these are based, as far as you can judge, on impeccable methodology and reasoning but the results are sufficiently different to have led to contrary conclusions. You have no reason to accept one view over the other. Your further reasoning will be clouded by doubts as to which of the two views you should use as a base. The approach is to admit that you, and the literature, are confused. For a reader, who is likely to be confused anyway, it is very helpful to know from the start that the material you are describing is in a 'grey' area. If you don't warn readers, they will think that their inability to come to a firm conclusion is your fault. Sometimes in these cases one of the most fruitful procedures is to try to devise and outline an experiment within the review that could be used to test which of the two views might be right.

By contrast to this example, you may sometimes find that the only information you have on a topic comes from one or several weak and questionable sources and you believe none of them. Once again you should be honest and admit that your further arguments on the topic are based on the best information available but which in fact you believe may not be reliable. This enables readers to make appropriate adjustments to their interpretation of your reasoning. Such honesty also ensures that your reputation remains untarnished should later and better experimentation demolish your tentative interpretation. You may ask, why make any interpretation at all if you believe the data to be of poor quality? The answer is that the advancement of science is a process of taking available information, poor though it may sometimes be, interpreting it, testing the interpretation and by so doing providing better information. As a reviewer your vital role in this chain is interpretation. If you do not play that role you might as well not write the review.

6

The Seminar or Conference Paper

A WELL-KNOWN scientist who is also a friend of mine was once asked to present an 'invited paper' at a scientific meeting. He confided to me that he would have to prepare two papers, one for presentation and one for publication in the proceedings of the meeting. It was not until later when I compared his excellent oral presentation with the equally excellent written article that I realised the amount of effort he had made to produce two different presentations of the same research material. By contrast, at the same meeting, in which the proceedings were available in advance, I watched a speaker reading from his written text. The audience amused themselves by following the text as a reading class might do at school. The only light relief came when the reader occasionally missed a line or a few words and became momentarily confused. The audience, of course, were able to read to themselves much more quickly than the speaker was able to read aloud and so had plenty of time in which to get bored. These two presentations were a good illustration of the wide difference between articles destined for reading and those destined for hearing.

Written Word and Spoken Word

The structure of spoken papers is controlled by the fact that speaking is a slow, relatively inefficient way of transferring information. In broadcasting there is a rule that says that it takes about three minutes to put across each new idea. If a presentation lasts for eight minutes, with one minute of introduction and one minute for conclusion, there is only room for two major ideas. Three minutes seems a long time to dwell on one idea and the structure of the spoken article has to be arranged accordingly. The speaker has to repeat the idea several times (not, however, in the same words) and to expand the presentation in order to give the audience the necessary three minutes to grasp the full implications of what he or she is talking about. You must realise when constructing your paper that, relative to you, the audience is uninformed on the details of your subject. Your task is to bring them up to date in a few minutes with something that you have been thinking about for a much longer time, maybe a working lifetime. No wonder they need some help. To give them this help and in fact to make yourself understood, you have to construct your presentation in a totally different form to that which you would use for a written article.

Length

Most inexperienced speakers overestimate the amount of material they can present in a given time. You can speak normally at about 100 words per minute. Slides or overheads are equivalent to about fifty words. The larger the auditorium the slower you must speak to be fully understood. Analysis of recorded speeches by famous persons to huge audiences has shown that the rate of delivery may be no more than fifty words per minute.

As a useful rule of thumb you can deliver 400 words each five minutes if you use illustrated material. This is approximately the number of words that occupy two double-spaced typewritten A4 pages.

In well-run conferences each chairperson has clear instructions to keep the session precisely to time. If one speaker is allowed to take too long it puts subsequent speakers at a disadvantage. So those in the chair generally warn speakers by a bell, a light, or a quick word that time is almost up and then cut them off in mid-sentence if necessary, should they attempt to exceed the allotted time. It is always distracting, sometimes amusing, and never informative, to listen to a speaker having a running battle with the chairperson while trying to finish the address after time is up. The details of the battle are sometimes long remembered whereas the hastened conclusions to the presentation are seldom assimilated.

The apparent ease and informality with which some speakers present their material and finish on time gives the impression that their talk has not been rehearsed. This is not usually true. The surest way to make certain your presentation will be within the time limit is to have a trial run in front of some patient colleagues. The next best is to closet yourself in the bathroom, the back shed, or the living room and rehearse it to yourself. In neither case should you forget the considerable time taken up while you present illustrative material. A presentation that goes even a few words over time must be carefully pruned because those few words could make the difference between an appropriate and dignified ending and a disorganised scramble. Don't forget, when preparing your article, that there is generally a question time at the end of most oral presentations. This can be used to fill in details should some member of the audience request it. Remember, too, that conferences and seminars are seldom confined to the sessions of formal presentation. There are generally many opportunities for people who have been interested by your talk to consult you privately on other aspects of your work. The essential requirement is for your paper to be stimulating and interesting in the first place. That is why your first priority must be to present your main ideas effectively.

Structure and content

Let us look at the major differences between the paper for reading and the paper for hearing. Table 3 suggests the approximate distribution of the components of the two types of article. The Introduction is relatively long in the oral presentation because it is necessary to put the audience in the picture. You cannot refer them to other people's work or to reviews. So the context in which the work was done must be clear and you must explain it before the audience can grasp the significance of the findings you wish to discuss. Then you present your Methods and your Results. In a spoken paper the methods must be pruned to the bare essentials by removing all distracting details. Information about the latitude of the location, the source of chemicals, the validation of the assays or the proprietary name of equipment may be important in a written paper but it merely diverts attention from your main ideas in a spoken presentation.

Table 3 Differences in the construction of papers destined for conferences and journals

Component	Paper for a seminar or conference	a journal
STRUCTURE		
Opening sentence	Sentence to make an impact	None
Introduction	40% of total (time)	5-10% of total (space)
Methods and Results	40% of total (time)	40-60% of total (space)
Discussion	20% of total (time)	30-60% of total (space)
Closing sentence	A clear resume of main point—similar or complementary to opening	None
SUBJECT MATTER		
Ideas	One every 3 minutes	No limit
Repetition	Highly desirable	Very little
Length	To finish just before time	As short as possible
Accessory material	Slides or overheads as reinforcement of text	Only those tables and figures that are relevant
Humour	Desirable but not essential	Undesirable
Grammar	1st and 2nd person used often	2nd person never used
Style	Conversational and simple	Formal and simple
References	The least possible	The required number for sound arguments
Acknowledgements	The least possible	Brief but adequate

Remembering the relative slowness with which ideas can be absorbed, you must be sure that those you focus attention on are the important ones. The results you report have to be those that emphasise the points you wish to make and must include no material likely to divert the listener on to another track. In planning the discussion, think of what you would most like the audience to remember at the end, about the material you presented. This must become the substance of your discussion with one good concluding point which makes an impact.

Presentation

Padding: how to repeat yourself without being dull

It is sometimes difficult for a writer who has conscientiously learned to write clear and concise scientific articles to present the same information effectively to an audience because of a conditioned avoidance of repetition. Yet to bombard an audience with a variety of new information in a very short time is fatal. The essence of making yourself understood is to present a single fact or idea in a variety of ways.

How can you vary your presentation?

You can vary your voice, vary your construction and vary your media.

1 Variety of voice

Lullabies are deliberately written to be presented in a monotone. Their purpose is to put people to sleep. It is a simple step of inductive logic to conclude that scientific papers presented in a monotone will also put people to sleep. This hypothesis has been tested and validated too often for you to waste time testing it again. One method of avoiding monotony is to vary the way you speak. You can practise dropping your voice at the end of sentences, modulating it through the sentence, and raising it when asking questions. Indeed, incorporating questions in your talk is another way of introducing variety.

2 Variety of construction

A film-maker who wishes to describe a series of events in, say, a single room, does so by breaking the whole presentation into a series of short sequences each seldom longer than about seven seconds. Close-up scenes are interspersed with long-distance shots and details with panoramas. The theme is the same each time but the approach keeps varying. This principle holds even if the total sequence in the room lasts for as long as an hour. If the scene for each segment takes longer than seven seconds viewers become bored. When the mood is one of excitement the elements become very short; to convey tranquillity they become long.

There is a strong similarity between film-making and oral presentation; the principles that hold an audience and induce it to follow a sequence of events are identical. When your presentation reaches its climax don't bury your most significant point in a long boring sentence. Present it to your listeners sharply and clearly and convey the excitement.

Assume, for example, that your research has resulted in some solid data that refute a previously accepted principle. Your paper describing these results would certainly wish to highlight this finding. Compare the impact of:

> In view of the results presented in this paper, and taking into account the large number of replications made over a wide variety of experimental conditions the author is forced to conclude that the long-accepted principle of reciprocal parity is not tenable under the circumstances described here.

with

> The principle of reciprocal parity doesn't work. It didn't work over twenty-two replications and it didn't work in any of the environments in which I tested it.

If you expound a principle, illustrate it with an example and some details in the very next sentence. We saw in the preceding example that the general statement that reciprocal parity does not work was immediately followed by a detailed statement of specific results.

Conversely, if you have described a series of detailed results or methods, change next to a general statement, either a summary or a relatively sweeping observation based on the previous details:

> 'Only ten per cent of insects in samples taken from three sites in the forest laid eggs in the two months after the winter solstice. Fifteen per cent of

those in samples taken from five heathland sites laid eggs in the same period. Thus, insects in this region have low reproductive activity in winter.'

The first two sentences give a number of figures and precise details. The presentation threatened to become too heavy if they had been followed by more of the same. Instead, the general statement that follows summarises the previous two sentences and allows the listeners time to catch up and assimilate the salvo of figures that has just been fired at them. If more detailed figures are to follow listeners will now be ready for them.

3 Variety of media

Here, at least, is where speakers have the advantage over writers. They have a wide variety of visual aids that they can use. Material which is unacceptable in journals is often highly effective to present to an audience. For example, if you are describing an experiment on a particular chemical molecule or using a particular apparatus, you can use one or more colour slides to describe it. An editor, ever mindful of space, could not allow such luxuries. Overheads can be used, fully prepared, or with overlays, or can be drawn during the talk to illustrate your points as you go. Remember, you are not restricted to dull figures and tables. Some relief from the concentrated data is not only possible, it is highly desirable.

There are, nonetheless, a number of precautions that you must take in using illustrative material. When people are suddenly presented with something new to read they seldom listen to what you are saying. This means that you might as well pause after a new slide or overhead appears on the screen. Or, if you must continue to speak, make sure that your first sentence does not contain a key statement. The 'dead period' can be used to orientate the listeners to the axes of a graph or the column and row headings of a table. The length of the 'dead period' is directly related to the amount of material in the illustration. This is why speakers are advised to present only the barest essentials in each slide or overhead. Moreover, when constructing the text for your presentation, it is essential to take account of the 'dead time'. Many a paper that fitted the time allowed when rehearsed in the bathroom has gone over time on the day because the 'dead time' associated with slides was not considered.

We looked earlier at the problem of presenting numbers in a talk. They tend to be dull and if there are too many, they sound alike. The best method of presenting numbers is by allowing the audience to read them. The commentary can then interpret the significance of the numbers rather than repeating them exactly. So, while pointing to the numbers we can use the expressions 'as many as half . . . '; 'almost half . . . '; 'only half . . . '; 'no more than half . . . ' to convey much more than the exact data and to set up the later discussion. If we have a slide showing that 1.7 per cent of rats had cancer after exposure to a treatment we might say ' . . . as high as 1.7 per cent'. If a crop had 1.7 per cent weeds we might say ' . . . practically no weeds'. Once again, the words convey much more than the exact figures.

Remember, too, that you will be helping the audience enormously by rounding numbers on slides or overheads to as high an order as possible. Most people are incapable of comparing and contrasting long strings of digits even in written articles where they can take their own time. In that case, we saw that you should attempt to round to just two significant digits except where this is unacceptable for reasons of precision. For illustrative material in a talk you should make no exceptions.

Delivery

When we talk of delivery we are talking about the art of presenting papers and seminars. We have all appreciated the experience of hearing, from time to time, a speaker who appeared particularly gifted. Some of us may not be capable of reaching the dramatic heights of others but we are all capable of producing a carefully constructed paper that can form the basis of a highly acceptable, if not brilliant, presentation. The style of writing must be adapted to suit delivery to a live audience. The conversational style of delivery has many advantages. The spoken word can be quite imprecise and may include fashionable jargon without being confusing because from other clues such as content, gesture, or tone of voice the audience can sort out the correct meaning. In the written word no such subtleties are possible.

When preparing your paper it is a good idea occasionally to take a rather pessimistic view of how the audience might be thinking at the time you will be presenting it. Imagine, for example, that you are the third speaker in the session after lunch and you have been preceded by two particularly boring speakers. The audience, despite its best intentions, is bored, sleepy and looking for distractions. You have been introduced by the chairperson and arrive at the speaker's desk. It would be unrealistic to think that, in that situation, the audience would now be waiting eagerly to gather every pearl you are about to cast before them. Instead they are perhaps thinking that your tie is crooked, you look nervous, the room is too hot, or the session is too long. Here is your first big hurdle—to unify their thoughts and to unify them so that everyone takes notice of what you are saying. You could make a spectacular opening by tripping over the microphone cord or knocking the water jug into the chairperson's lap but this would be a hard act to sustain for the rest of the talk. Words are your main equipment and your opening sentence is crucial. It must make an impact and at the same time make people wish to hear more. For example, an opening such as:

> Bloggs et al. (1992) in extensive studies into partial segregation of myoepithelial cells concluded that little segregation is evident in non-pathological subjects . . .

would turn off all but the most ardent listener from the very start. Another approach that is almost as dull is to recite the title of your talk.

> The title of my talk this afternoon is the effect of culture medium on the separation of myoepithelial cells.

If the person chairing the meeting is doing a proper job the audience will already know that.

How much more animating (at least to an enthusiast of myoepithelial cells) is:

> This afternoon I am going to show you that myoepithelial cells can be separated easily if you choose the correct medium.

Apart from giving more information, this opening sentence tells the listener what to expect. It is a resume of the whole paper in one sentence. But have you revealed too much of your talk by adopting this approach? Not at all. Your function is to get across a message, not to have secrets. The adage 'Tell them what you are going to say, then say it, and then tell them what you have said' works wonders in presenting scientific papers. For our purpose the adage can be translated: Make an arresting beginning containing a 'micro-summary', then use the body of your paper to present evidence and convince the audience, and then round off with a conclusion containing the 'take-home-message'. Not surprisingly, the 'take-home-message' is exactly the same as the message you opened with. So, if you think that you have problems in deciding how to finish off a talk, your problem is solved.

Having gained the attention of all of the people likely to be interested, how do you keep it? We have already seen the value of variety. In addition, it is essential that the audience feels involved with the work you are presenting. This requires a very important departure in writing style from that which we discussed for the journal article. A conversational style, contrary to some views, does not reduce the scientific merit of the paper. The magic word that helps produce a conversational style and which should be used as often as seems sensible is the word 'you'. Whenever it is used the listener feels a participant in the paper.

'You may wonder why we used . . . '

'If you look at the two numbers on the right hand side of the slide you will notice . . . '

'The slope of the line is not as steep as you might expect . . . '

Each time you use the word 'you' the listeners, whether they like it or not, are compelled to make certain that they are not being maligned, misinterpreted or otherwise taken in vain and in so doing automatically pay attention. To a lesser extent the word 'I' (not 'we') and the active rather than the passive voice achieve a similar result. They involve the speaker with the substance of the talk:

'I couldn't get two sets of data . . . '

instead of:

'There were two missing sets of data . . . '

'I centrifuged the material at 2000 revolutions per minute.'

instead of:

'The material was centrifuged at 2000 revolutions per minute.'

This personal style stirs the imagination of your listeners so they can picture you working your way through your experiment.

Another way of retaining attention is to include some humour in your talk. It lightens a heavy session and if you can successfully incorporate a joke in the early part of your talk you will keep at least some people's attention if only because they are waiting for another. Unfortunately, not everyone tells jokes effectively and laboured humour is worse than none at all. Anecdotes that commence with 'Did you hear the one about . . . ' or 'That reminds me of the barmaid with one eye . . . ' should be avoided. Successful speakers at scientific meetings develop their punch lines as very slight variations in the serious text and rely on unexpected turns of phrase rather than well-defined jokes. In this way the humour wastes little time and gains a great deal of attention—if it works. If it doesn't work, as occasionally happens with dull or sleepy audiences, you can continue on without even revealing that you expected a response. If you have set yourself up with an introduction like 'Did you hear the one about . . . ' you can't avoid being acutely embarrassed when there is no response to the punch line.

The following is a well-prepared seminar given by Professor R.J. Moir. It is not only useful to see how he incorporates many of the features we have just looked at but, because of its content, it can add to your armoury of seminar techniques. It is also set out in the form that he habitually uses. The left-hand margin has key words to which he can refer very briefly but, if he forgets a detail, he can re-establish his train of thought by reading from the complete text.

A Seminar on Seminars by Professor R.J. Moir

(First presented 1 July 1974; this is a long while ago but the principles have not changed.)

SEMINAR SEMINARIUM
A SEED PLOT

A meeting of students particularly to discuss research
formally or informally

SYMPOSIUM	syn. together + posis a drinking
	There are so many 'seminars' these days that one goes either for information, for a sense of duty or, more rarely, for the pleasure of hearing about an interesting subject and all the better if it is an interested, enthusiastic speaker.
ENJOYMENT	If, you the speaker, can communicate your enjoyment and interest, or the thrill of your discoveries, the audience will soon share that enjoyment.
REHEARSAL	Provided you have rehearsed your communication, you too should enjoy it. You have the first advantage, too, that you are talking about your work—you are the expert and you should know a great deal about it.
NERVOUSNESS	To be nervous before an audience is to be natural: the audience usually is not, so you have
ADRENALIN	the advantage of a little more adrenalin to sharpen your wits and to be that little bit
LARGER THAN LIFE	larger than life, which at a distance of three, ten or more metres, makes you appear more natural.
BREATHING	A chestful of air is a wonderful antidote to butterflies in the stomach so that a deep breath is a helpful start to a seminar.
ART OF COMMUNICATION	Scientists should be versed in the art of communication, written and spoken, privately or publicly: even the best work is nothing unless you can read about it or hear about it. It is the exceptional person who can give a seminar
WRITE OUT FIRST	without writing it out first and it would seem to be elementary that this is a must. Writing crystalises your thinking and also gives you a guide to time.
INTRODUCTION	A brief introduction is often helpful, particularly when you are speaking to an unknown audience

or to an audience who are not all specialists in your field. Usually a few minutes is enough and it is the first danger area as far as audience receptivity is concerned.

EDIT OR CUT

Be prepared to edit it down or out if need be. If you have cast your talk at too high or too

LEVEL

particular or too general a level you may have to give an entirely different presentation based on audience reaction. You can afford to lose some of the audience, but not the lot.

JOKES

Many speakers introduce their talk with a witty joke: it depends how good you are at telling a joke and how relevant it is. The one about the actress and the butcher never goes over at a scientific seminar. Some people end up being invited to give the conference after-dinner talk if they are good at this sort of thing, but never giving a scientific paper at the conference.

HYPOTHESES
PRECONCEIVED IDEAS
FINDINGS
INTERPRETATION

Develop your topic: the development of your hypothesis, the experimental regimes and your difficulties, leading to your exciting findings and then your ideas about their interpretation. I find it difficult to suggest how to write because each individual should generate and

STYLE

develop their own style. Style is important because it tends to convey some of the author's

PHILOSOPHY AND
FACTS

philosophy as well as his or her facts: a recital of facts, however important, rarely stimulates.

PRESENTATION

Having written your seminar, how do you present it?

READING
MUST WRITE TO BE
LISTENED TO, NOT
READ

You may read it: very few people can do justice to themselves this way. Even good readers do not project the subject well unless the material has been written for speaking.

WRITING 25wpm

I can only write about 100 letters or twenty-five words in a minute and although

TALKING 100 wpm

I am a slow speaker I speak at about 100 words a minute and although a slow thinker,

THINKING
600–1000 wpm

I probably think at six to ten times that rate. So writing tends to be slightly stilted.

WRITE CLEARLY

If you do read your paper, write it out clearly so that you do not have to turn the paper to read comments in the margin. If you have it

TYPED—CHECKED

typed, check it before you present it. I have seen some quite high-powered scientists put off balance because a typist skipped a line; and that can wreck a tight argument. You are giving the paper so it is no good blaming the typist.

FOUNTAIN PEN
BALL POINT

If you do write it out yourself, use a fountain pen rather than a ball point pen. A ball point stretches the paper and curls it—often sufficiently to raise it clear of the ledge at the bottom of the podium, so that when you turn to explain a point on a slide, or there is a slight breeze or an earthquake, or you brush it with your sleeve, your beautiful notes flutter gracefully one after another onto the floor. Because they have different curves they go in different directions. Your discomfort at finding the wrong page is heightened by the embarrassed murmurs or titters of delight from the audience, by the fact that one page looped the loop to the front, that you forgot to number the pages, and so on. This can and does happen. If it does, one

AUTO TO MANUAL

merely changes from automatic to manual and proceeds as though nothing had happened.

AIDE MEMOIRE

I prefer listening to a speaker who is making use of what memory aids he chooses. He may use his full paper with underlined headings, the headings themselves as notes or on cards or, best of all, straight presentation from memory and knowledge. One expects a professional actor

ACTOR
PERFECTION
THROUGH
REHEARSAL

to be able to present his lines in a play without prompting and with the minimum of paraphrase: they are too concerned about their professional status and so should you be: actors attain their perfection by rehearsal; so should you.

ON SPEAKING

If you read, hold your paper up or, better still, read a small subsection at a time. At the end of each subsection drop your head, read in silence, compose your thoughts, lift your head and continue the lecture.

VOICE PROJECTION · Lifting your chin tends to increase the pitch of your voice so that it projects better. Nothing is more frustrating than not being able to hear MUMBLING · what is being said. Mumbling at the desk, and BACK TO AUDIENCE · deliberately turning your back to the audience to talk to the screen where your slides are projected is unforgivable and unprofessional.

MICROPHONES · If you have a fixed microphone, maintain the correct aspect to it. Microphones are often extremely directional and very sensitive. They introduce another dimension and require a technique of their own. If there is a mike, adjust it to your height and don't try to put a curve in your neck. Don't rely on the microphone to correct your mistakes of presentation. Speak in your normal voice as if to someone in the opposite corner of the room. Let the fellow who twiddles the knobs sort out the correct level for the theatre.

SPEAK TO AUDIENCE · Speak to members of the audience: as they have INDIVIDUALLY · had the courtesy or desire to come and hear you, speak to them individually.

AUDIENCE RESPONSE · The response of the audience can help you in your presentation. I find that speaking at the row about two-thirds of the way up an auditorium gives me about the best aspect in terms of my own voice production. Departures from this attitude such as reading writing on the board are best done in silence. Well-organised silences are as telling as well-presented work. If the audience is happy I am happy too. You can see whether they can hear, whether they are interested, or going to sleep.

OBJECTIVE—USE · You have to work on the audience: remember AUDIENCE · you are talking 100 wpm. They can think at THINK POWER · 600–1000 wpm and your job is to get them to use up their excess think power on your subject and on your work; if they start using any on last night's assignation you have lost, although some audiences are good actors: if they stop thinking it is time to re-write your talk or conclude it.

SLIDES	Presentation slides, charts, etc. Make sure the slides are in the proper order and are correct. Too many authors find mistakes when visual aids are used for the first time.
CONTENT	Do not have too much on the slides.
TIME TO SEE	Do not flash them on and off. Subliminal information is worse than having lights out all the time.
TOO MANY	Do not have too many slides, particularly in a short paper, as it takes a little time to point out the salient features. Slides take a lot longer to present than you think.
IMPORTANT POINT	Emphasise only the important point. The audience is not dumb . . .
TALK TO AUDIENCE	and talk to the audience, not to the screen. Sometimes speakers have to present their own slides.
MECHANICS	Go early and have a practice run.
TALKING COCKPITS	Some of the complicated talking cockpits in modern lecture rooms are hazardous places and it is quite off-putting to press the wrong button and find that the screen disappears.
MALFUNCTION	Malfunction is not impossible and is even a probability. Always have on hand the Table from which the slides are made so that as soon as malfunction occurs you can move to
CHALKBOARD	a chalkboard and write up the main figures large enough to be read at the back of the auditorium.
INDICATE CONCLUSION	Clearly and definitely indicate your conclusion either by statement or preferably by resume, and do this before chairperson turns off the mike.
FINIS	I have heard speakers say 'I will now answer your questions.' I consider this presumptuous. For one thing there might not be any questions and it is invariably the chairperson's prerogative to invite questions.
TIMING	The timing of papers is a very important part of the good pro's presentation. Prepare your paper to a time and do not forget that slides take time to present. In preparing your paper and in rehearsing it, give some thought to

EDITING ON
YOUR FEET

'in presentation' editing. Thinking on your feet during presentation is something to learn; it is part of the professional presentation, and much easier if you are not reading the material.

QUESTIONS
BRIEFLY REPEAT

Questions: If someone has screwed up enough courage to ask a question, give a brief answer and not another paper. Outline the question if the auditorium is a large one, because the questioner does not have a microphone and some may not have heard. An answer without a question is silly. You should cope with most questions, but difficulties do arise.

BE HONEST

If you don't know the answer, say 'I do not know' and no more. You are seldom criticised for a statement as honest as that. However, you may know that a colleague knows something of the topic so refer to chairperson. 'My colleague, Mr X, may be able to answer that, Mr /Madame Chair'.

PASS THE BUCK

TOUGH QUESTIONS

Then there is the really tough question that takes a little time to think about. A stunned silence doesn't help you or the audience, nor does a string of 'ums' and 'ers', however well modulated; the standard gambit here is to say, for example 'Yes, Mr Chairman, that is a most interesting question' (three seconds). 'I am glad the speaker raised it' (another two seconds) and, if you wish to indulge in a little sciencemanship, you add 'it has intrigued me for some time'. This has given you ten seconds or a least 100 thought words to get going.

SCIENCEMANSHIP

DISCURSIVE
QUESTION

There is also the discursive question, part egocentric discussion (on the part of the questioner of course) interspersed with vague ill-defined questions. The only gambit here is to ask the speaker politely to repeat what he said, whereupon he usually asks only one simple, easily answered question.

MULTI-PART
QUESTIONS

There are often, however, legitimate multi-part questions; it is always advisable to jot down key words to the various parts so that you answer them in the sequence they are asked. Nine times out of ten, speakers answer the last question first and then have to ask questioners to repeat their earlier ones. Five times out of ten the questioners have forgotten them too. However, it is often advantageous to rearrange the order of the questions to suit your train of thought.

UNFAIR QUESTIONS AND COMMENTS
There are, too, unfair questions which can be dealt with by suggesting discussion outside the meeting, and unfair comments which are hard to deal with, but like people talking in a theatre or laughing and clapping at the wrong time these have to be contended with.

YOU THE SPEAKER
Then, finally, there is the speaker. Remember you are on show too. Anything that distracts the audience should be avoided and this takes a little discipline.

DEPORTMENT
What do you do with your hands? You use them to turn pages and handle pointers.

HANDS
Putting both in your trouser pockets doesn't look too good, nor leaning on them on the podium. If you are nervous, holding the edges of the podium helps immensely, and your hands are ready to use and to emphasise a point.

PACING
Walking up and down like a caged lion is also distracting and should be avoided. Most of the time only half the audience can hear you anyway. The microphone cord can get tangled in your legs and it is even worse if you carry the slide-changing device with you and emphasise your point by pressing the trigger as I once saw some-one do. When he went to give us slide No. 2 we had already seen 20 slides go whizzing by:

CONCLUSION
In conclusion, the presentation of scientific material is an art and an essential part of your professional armoury.

It has been said that an experienced speaker is one who goes on making the same mistakes with increasing confidence. A great expert speaking with humility makes one of the most impressive of all communications, and I hope you attain that standard; you need practice and the help of friends and critics.

7

Student Practicals and Reviews

S CIENCE COURSES usually involve a large proportion of practical training. This may come in a variety of forms such as formal practical classes, field excursions or assignments based on reading published scientific material, and they all involve the collection of data from experiments, surveys or written material, together with presentation and discussion for assessment.

There are three reasons why this training is included in scientifically-orientated courses: to develop a familiarity with the language and equipment of the branch of science being studied; to foster and develop skills in the technology believed to be essential for a graduate; and, most important, to develop an approach to the collection, collation and reporting of data.

We are concerned here with the techniques of reporting clearly, logically and scientifically. It is obvious that for prospective research workers, writing practical reports in a format like that used in writing research papers provides valuable practice for a future vocation. Perhaps it is not so obvious that this same training will be equally valuable for scientists who will not do research after graduation. All scientists whether they are advisers, technologists, consultants or private practicians have to read research results to remain in touch with their fields. An appreciation of what goes into a scientific article can best be acquired through practice and criticism during the undergraduate course. For this reason, practical exercises and the consequent practical reports are an important part of undergraduate training. The skills needed are precisely those we have already encountered in the sections on scientific articles and reviews. In this section, we will concentrate on ways of adapting the principles discussed earlier to the construction of assignments. These can be one of three types:

- Practical reports in laboratory based subjects;
- Reports in non-laboratory subjects; and
- Essays and reviews of the literature.

1 Practical Reports in Laboratory-based Subjects

Many research projects can take as much as six to twelve months to complete but most practical classes last for only one or two afternoons. Practical classes are therefore limited to only a few of the elements of typical research projects. Most of the thinking behind the practical class, the design of the experimental work, and the preparation of reagents and equipment is usually already done by the supervisors when the students first begin to think about the work they will do. So the practical work itself will be to some extent artificial. The work will not be new to the world and the experiment may be unrealistically small with numbers of samples and treatments being constrained to the limits of time and space available.

Nevertheless, when it comes to writing up the work the final reports should contain most of the elements seen in good scientific papers. After all, if you are a student the work will be new to you, and it will be worthwhile as a training exercise to present the work in the form of an experiment as if you had been involved in all its aspects. Your supervisors have designed the practicals in this spirit and, unless there are special reasons for doing otherwise, they will assess your written report with this in mind.

The components will therefore be those of a standard scientific paper:

- Title
- Summary
- Introduction
- Materials and Methods
- Results
- Discussion
- Acknowledgements
- References

You may be specifically asked not to include certain components such as, for example, the Summary and the Acknowledgements. But, if no special instructions are given, you should assume that specific training in scientific writing is an integral part of the practical class and that you will be expected to present all of the components.

The Introduction

We saw earlier that an Introduction to a scientific article should contain an hypothesis preceded by a series of logical statements that make that hypothesis a sensible thing to have tested. In the instructions for the practical class you may have been given the hypothesis but this is not always the case. The hypothesis is sometimes deliberately omitted or merely implied. The purpose of this is to encourage you to read enough about the subject to take the simple statement of aims or objectives which you have been given and develop it further to formulate an hypothesis.

For example, if the practical class notes said:

> The aim of today's experiment is to cook meat from old and young sheep—mutton and lamb—at three temperatures 60°, 75°, and 90°, and examine the degree of loss of moisture.

You would be expected to do much more than repeat this aim as your Introduction. You would be expected to think and read about the subject to come up with one or more hypotheses which in this case might be:

> The experiment was a test of the hypotheses that meat from old animals loses less moisture than meat from young animals and that the degree of loss depends on the temperature of cooking.

You would now have to present reasons why you have these expectations (or hypotheses) about the outcome of the experiment. You might explain that water holding capacity is related to tenderness (with a reference) and tenderness is related to age (with another reference). Then you would need to show (with references) why you think that more water would be lost at high, than at low temperatures.

When you have done this you will have constructed your first paragraph which, together with the hypothesis, will complete a good Introduction.

A second example:

The object of the experiment is to study the percentage germination of rice and wheat seeds in growth media having different concentrations of salt.

The hypothesis might be:

Rice seeds are more tolerant to salt than are wheat seeds.

Rice can survive much higher waterlogging than wheat and this should give it higher tolerance to salt. Your Introduction should develop this point and any other you believe relevant to the hypothesis and support them with references wherever possible The final effect will be to produce an Introduction very different from, and far more meaningful than the mere instructions that you were originally given. Such an Introduction will also give you a sound basis for constructing the rest of the report.

The Materials and Methods

The methodology is usually given to you in detail in most practical laboratory experiments. Nevertheless, you must ask yourself whether the instructions were as clear as they could have been. You are, after all, a good person to judge this because you have had to do the experiment by following the instructions. If you had some trouble in understanding them or carrying out the technique you should take the opportunity to clarify the written material in your own version. Sometimes you get important information, from 'Prelabs' or discussion with demonstrators or fellow students which was not included in the written material. Your report will be the more impressive if this information is incorporated logically into your own Materials and Methods section. If you do it well you will gain the immediate benefit of impressing your supervisor and maybe the longer term benefit of helping future classes if your suggestions are adopted into the official notes. Furthermore, you may need to refer later to changes in the methods you used from those stated, to explain some aspects of your results.

The Results

Results should, of course, be reported correctly and faithfully but this does not mean that you should present all of your raw results. One of the most common faults in practical reports is that the results are presented in the form of a direct transcription from the laboratory notebook. Your chances of achieving good marks will rise substantially if you can show that you have thought about the raw information and have processed it into meaningful tables, graphs, or diagrams which smooth the way for a logical Discussion section. In most cases a mean and its standard error are more valuable than the large number of individual records from which they may have been derived. You are not hiding results or taking short cuts by leaving out the individual records. In assessing your work, supervisors are looking for your ability to discriminate and interpret. Filling pages with raw data does not demonstrate either of these skills. Remember that the hypothesis you presented in the Introduction section should be the guide that directs the way you process your data. The final presentation should be in a form that most easily allows you in the Discussion to accept or reject the hypothesis.

Tables, graphs and diagrams in practical reports, as in scientific papers, should be fully and correctly titled and be 'self supporting'. Similarly the text should be readable without the tables and graphs but, of course, complemented by them.

The Discussion

You should not have too much trouble in separating the major points of discussion from the minor ones, because most practical exercises are relatively simple. The main topic for discussion should be whether or not the hypothesis is supported. This will occupy the opening, and possibly the only, paragraph of your Discussion. If there are other interesting points arising from your results that are not related to the discussion of the hypothesis they should be presented in succeeding paragraphs.

Sometimes discussions of practical exercises can be difficult because the data are not as clear as they could be. An exercise lasting for only a few hours usually has imperfections. For example, there may be insufficient results to ensure statistical significance of differences between treatment means, or short cuts may have to be made in the design which could confound the results. Such inadequacies would not be tolerated in a research project destined for publication in a journal. In order to make a worthwhile discussion of a practical exercise it may be necessary to take some licence. Remember, you are involved in a training exercise in writing and to make the exercise worthwhile you may have to acknowledge that you were not able to fulfil the technical aspects as well as you would have liked. Perhaps you will have to accept a lower order of statistical probability than five per cent or admit that there could be a confounding of the results.

If you are forced into this position the important thing is to acknowledge the problem you face. In so doing you should describe clearly how the experiment might have been done better had you had more time or materials than was possible in the practical class. Then, and only then, should you go ahead and discuss your results as if they were a little more convincing than they are.

Occasionally, you may have reason to doubt some of your results owing, for example, to your lack of experience in the necessary techniques. Ordinarily, you would repeat the part of the work in doubt, but this is generally not possible in practical classes because of lack of time. You may decide to leave out those doubtful results in order to make your conclusions clearer. Take care to explain fully your reasons for omitting these data. Without such explanation it will appear that you were simply trying to make your results fit preconceived conclusions.

The Acknowledgements

Normally there will be no acknowledgements in a practical report. Lavish praise of your supervisor is unlikely to get you extra marks. But, if you have used extra data from another source such as a colleague or another practical class group, then this is the section in which to say so.

The References

References should be handled exactly as for a paper for a scientific journal. They should be exact, correctly abbreviated according to a specified format, and cited in both the text and the References section.

Papers and books that have been consulted as background reading but which have not been cited in the report itself should not be included.

Not only are practical periods too short for many experimental exercises, but so too is the time available for the writing of reports (although some individuals seem to have more than others). It is seldom possible to write three or four drafts of practical reports. Nevertheless the hallmarks of scientists are their precision, clarity, and, to a lesser extent, brevity. It is simply not possible to give the impression of having any of these attributes by handing in an unchecked first draft of a practical report. Reading your own report to check for obvious errors will be time well spent. A second step, requiring no more time and often yielding a similar degree of improvement, is to ask a class-mate to read your report and comment on it. Perhaps you can do the same for them. This is not cheating. A practical report is not an exam paper, although it may contribute to your final marks. To use the criticism of colleagues to produce a scientifically acceptable report (providing of course it is you who will make the eventual corrections and changes!) is eminently scientific and laudable. Not only that, it is an excellent way of learning.

2 Reports in non-laboratory based subjects

Where the subject of the report is not based on a standard scientific investigation but on gathering information from the literature, from newspapers, or from interviews, the structure may need to be modified. The principles of writing however, remain the same. Readers needs to know where you are leading them.

In the science-based report you used the hypothesis to create an expectation of what you were looking for, and a yardstick with which to compare the results you found. If you are not going to have a hypothesis, then you must substitute another method for creating expectation. When you write an introduction you should try to give readers mental images and questions that will guide them through the rest of your report.

Paradoxically, the most effective way of doing this is to present your main conclusions in your introductory paragraph. These give a sense of logic to the way that you gathered your data and developed your arguments. The fact that you will re-present the same conclusions at the end of the report does not detract from their benefit at the beginning. In fact, just as with hypotheses, they have the important function of 'rounding off' the report and reiterating its most weighty outcomes.

Let us suppose that you have an assignment to present a report on the impact of urban bicycle paths on the attitudes of city dwellers to exercise and transportation.

Compare these two introductions:

> The use of the bicycle is a time honoured way of gaining non-stressful relaxation and exercise while at the same time providing low cost transportation. Unfortunately, bicycles and cars are incompatible in the inner city environment because of the physical danger to cyclists and the perception of chemical danger through pollution from automobile emissions. A solution to this problem is to separate the two forms of traffic by constructing dedicated bicycle paths. This course of action has been taken up in recent years by many civil authorities including our own, and we are now in a position to assess how successful this has been in improving the life style in our town. This report presents preliminary conclusions about the effectiveness of the bicycle path program in Bloggsville.

This is a 'standard' introductory paragraph which sets the scene, introduces the topic of the report, and builds up to say what the report is going to address. Most people would be happy that it is doing its job. But compare its impact with the more direct introduction.

> This report shows that only 10% of the residents of Bloggsville make use of the new bicycle paths in the town. A survey of residents revealed that they perceive that the routes are poorly designed for the needs of urban commuters. It is proposed that the Town Authorities campaign to advertise the useful circuits available on the bicycle paths and the advantages of health and safety to increase their use and popularity.

Now ask yourself:

- How many sentences in each introduction?
- How many relevant questions have each of them posed?
- How many irrelevant images have each of them posed?

The second introduction has provoked a much greater sense of expectation than the first and the reader will be more at ease with the text that follows. The general background information in the first paragraph is still important and, indeed, almost all of it could be presented as the second paragraph of the report. The difference now is that the reader is absorbing this background information in full knowledge of where it is leading. Less important (but every bit as relevant), the writer now has a clear basis for deciding what is acceptable background material and what is not. This is exactly the same basis that helps decide what methodology will be included, what data will be included and how they will be presented, and even how the discussions and conclusions will be handled.

If reports are read by busy people—and isn't that always the case?—then an opening paragraph like the second one will ensure that they will get a clear message from the briefest of glances at the report. By contrast, the first paragraph will convey little more than the subject to be addressed.

3 Reviews and Essays

Another form of practical assignment is a literature review where the research is carried out in a library, tracing and studying original references.

Reviews by undergraduates cannot be expected to be as comprehensive as good reviews by experienced scientists. Good reviews contain a substantial proportion of new thinking based on the cited work and a lot of this new thinking is the result of years of experience in the field of study. Nevertheless, inexperienced reviewers must still try to inject some of their own thoughts into their work. Simple lists of data and references do not constitute a review. Not only that, they are dull—dull to write and even duller to read.

How can we avoid writing catalogue-type reviews? Much research is characterised by the variability between individuals, means, and populations. Conclusions and generalisations about the one subject made from several sets of data often differ, simply because data vary enough to suggest different interpretations. In your reading you will repeatedly encounter differing points of view or data that appear to conflict. You may not have the experience of the authors whose work you are reading but you do have the advantage of hindsight. You can look at work done several years ago in the light of more recent knowledge; the original author might have had to assess his or her own work in relative isolation. This gives you the opportunity to reassess the conclusions and to compare them with those of later or contemporary workers. You may even see results which the author thought little of at the time and failed to discuss adequately. These results may appear to you to be very significant in the light of later work. If so, you have an excellent chance to make an original contribution by drawing attention to its significance. Thus, when reading the literature for your review you should be alert for opportunities to:

- Compare results and conclusions by different authors.
- Contrast results that appear to lead to different conclusions.
- Reassess results in the light of new information that might not have been available to the original authors.

In each case you should be prepared to make a statement that rationalises, explains, confirms, or refutes the work you are going to cite in your review. Not all information is amenable to such analysis but if you can find just a few examples it will lift the quality and impact of your review immeasurably. Assessors are attracted by original thinking. Even if you are not always considered to be correct you will get good marks provided that your attempt has been characterised by thoroughness and logical thinking.

The key to a good research review is an effective introductory paragraph. Here is where the reader is told what to expect and—just as important—what not to expect from the review. In most cases the title you have been given will allow some latitude in scope and emphasis so that you must define for the reader the exact limits of the material you will be covering. If the reader's interpretation of the title differs from yours he or she may expect to find information that will not be in your review. By carefully defining from the start what you are going to discuss you can avoid disappointing such readers. You can help them even further by indicating the order in which you plan to discuss the information.

Thus the first paragraph in your completed review has the essential role of guiding the reader. Because of this, you may find that you cannot construct it satisfactorily until you have completed the main body of the material, or at least a detailed outline of it.

The concluding paragraph, too, has an important role. In it you can emphasise the main points you feel you have made in the review. If these points have already been foreshadowed in the Introduction the review will have a sense of coherence and completeness. The impact of your review will be very much greater if these main points to which you refer in the concluding paragraph are the ones that you yourself developed rather than those taken from cited references.

As before, editing, re-reading and re-drafting are essential if you want to hand in a review of which you are proud and which may result in good marks. Generally a review will have more references than a practical report and these, as always, must be checked for accuracy.

You should make certain that you have not, consciously or unconsciously, been guilty of plagiarism—the uncited use of words or work of others. Sometimes in making notes you may write down, without change, expressions used by authors, and by chance incorporate them later in your own version without attributing them correctly. If such expressions are essential to your essay you must use quotation marks and cite the author, and you must be very careful that the quotation is accurate. Quotations are often very useful where a particular explanation has been well expressed or where an author appears to have been prophetic. It is nonetheless a good idea not to have too many quotations of this sort. They detract from your earlier efforts of originality by emphasising that you are not even using your own words, much less your own thoughts.

The essential ingredient in a review, then, is the expression of your own ideas. It makes the difference between a dull catalogue of facts and an interesting account of a field of study. Your overall aim should be to inject evidence of your own clear thinking into the introduction, the body and the concluding paragraph. If you succeed, your potential as a scientist will be unquestioned.

8

The Thesis

ALL OF THOSE elements that distinguish good scientific articles and reviews are found in good theses. The aspect above all others that characterises theses is their length and it is this which often causes problems. Theses are the written evidence of sustained research that has taken from one to maybe five or six years. They generally contain an obligatory review of the literature as well as the research material which in itself is often sufficient for several research papers. The problem is, as always, one of coherence. We have looked closely at how an article for a journal can be made coherent and therefore easy to read by using the development and testing of an hypothesis as a theme. Unification of the thesis uses the same technique, as we shall now see.

Form and Layout of a Thesis

A typical thesis consists of:

1 General Introduction
2 Review of the Literature
3 Materials and Methods
4 A series of chapters each containing one or several related experiments
5 General Discussion (and conclusions)
6 Bibliography
7 Summary

Added to this may be small sections for acknowledgements, indexes, appendices, statutory declarations, and other material which may be demanded by the institution supervising the thesis. Let us consider the various sections in more detail.

The General Introduction

The purpose of this section, apart from providing a background for the material to follow, is to set up what we can call the unifying hypothesis. This is different from the type of hypothesis we have been concerned with so far in that it is less specific, but it does provide a reasoned argument that justifies doing the series of experiments that follow. Later on there will be further hypotheses for each experiment. These we can call specific hypotheses. Let us examine some samples of how the unifying hypothesis works in practice.

G.J. Sawyer presented a thesis at the University of Western Australia in 1977. His General Introduction contained the following facts:

1 In general, ewes mate every sixteen to seventeen days during the breeding season until they become pregnant. After they become pregnant they no longer mate and this can be a good way of identifying which sheep are pregnant at the end of the mating period.

2 In his region, many ewes which appeared pregnant by this criterion did not have a lamb.

3 The temperatures during the mating season were always extremely high.

4 High temperatures are known to affect the survival of embryos in other species.

Using these facts, he induced an hypothesis that the failure of apparent pregnancies in sheep in his region was due to the high temperatures around the time of mating.

To test this hypothesis he needed to do a number of experiments. In one experiment he developed a test to enable him to determine exactly when an embryo died. In another, he looked for a critical time of heating which would kill embryos. In another, he tried various temperatures and so on. Each of these experiments tested a specific hypothesis. The results of each of them were pieced together finally to enable him to test the original, unifying hypothesis. In this way he brought together the report of his work under the umbrella of his unifying hypothesis so that at every stage it had purpose and direction.

It is interesting to note that in his conclusion he was able to reject the unifying hypothesis and come up with a completely new hypothesis based on his results. This made no difference to the unity of his thesis or to the ease with which he was able to write it. More importantly, it in no way reduced the coherency of the thesis for the reader.

A second example is that of a student who developed his thesis from this information:

1 A certain forest tree species was being attacked and killed by a fungus.
2 The damage was invariably found in trees in low-lying and wetter areas of the forest.
3 The tree species was found in association with different understorey species depending on the incidence of fires and other random causes.
4 In certain plant associations the trees remained unaffected even when soil conditions seemed favourable for the disease.

His hypothesis was that the disease could be controlled by encouraging certain plant species which would be unfavourable to the fungus to grow in association with the trees. Once again, to test that hypothesis he needed to carry out a whole series of experiments each testing a specific hypothesis.

The essential feature from both examples is that the purpose of the thesis is clear to readers from the very beginning. They can thus progressively make personal assessments of how the results meet the objectives of the thesis. In other words, the thesis is unified for the reader by the hypothesis.

The construction of the General Introduction is similar to that of the Introduction of a scientific paper which we examined earlier. The unifying hypothesis is carefully developed and this is the subject of the last part of the General Introduction. Then the first part is constructed from a logical sequence of information that make the hypothesis a sensible thing to test. The available data and information can be sifted easily and rejected according to whether or not they are necessary to meet this objective.

The Review of the Literature

The literature review for a thesis is only slightly different from that which might be published as a scientific article. Its objective is more clearly prescribed, in that it is meant to act as a base for the experimental section of the thesis. It is essential that all of the literature leading to the development of the hypotheses for each experiment should be covered. Some universities and some supervisors demand more than this and require that the Review of the Literature act as a demonstration that students have read widely, even beyond their immediate fields, if these are considered too narrow.

Regardless of the scope of the Review of the Literature, those parts that are directly related to the thesis should take the reader almost, but not quite, to the point where the specific hypotheses are presented. In this way the formal introductions to each of the experiments later on can be developed in the minimum of space.

Where there are no definite rules about the breadth of coverage in the review it is sometimes difficult to decide how to limit the amount of subject matter. A successful strategy is to assemble all material which will lead to the development of each hypothesis to be tested later in the thesis. Sometimes this material overlaps and the material for several hypotheses can be amalgamated into one section. When this is done you may find that there are distinct gaps between sections. New material may have to be introduced to unite all of the sections into a coherent structure. For example, let us assume that there are five environmental factors that predispose a crop to attack by insects and your thesis presents a detailed study of two of them. It is highly likely that a worthwhile literature review would have to discuss the other three factors at least to some degree in order to achieve balance. What is important is that all of the material introduced into the literature review has a purpose; either to develop arguments for use in the experiments to be described later, or to unify these arguments.

The Materials and Methods

A thesis often describes a number of experiments but these generally have several features in common. They may have been carried out in the same region with the same population of patients or on the same type of soil; they may have used the same micro-organisms, the same chemical analyses. In other words, they have in common most of the 'materials' part of the 'materials and methods'. To present each experiment with a complete description each time would be both boring and distracting. Instead, it is common to include a chapter that gathers together the materials and techniques used in most of the experiments. This has two advantages. It avoids repetition and it clears the way for the results of related experiments to be presented uninterrupted by long tracts of methodology. This chapter may also contain validation of methods used, even if in some cases they may involve small test experiments.

Each separate experiment will still have its unique features, the most notable being its experimental procedure, or the 'methods' part of the 'materials and methods'. Under the heading Experimental Procedure comes a description of the way in which the experiment was carried out. Details of techniques and methodology need not be included because they have already been covered in the special chapter for Materials and Methods.

The experimental procedure section for each individual experiment is best placed just before the results for that experiment. It puts the experiment in perspective but is compact enough not to act as a visual and mental barrier between related experiments.

The Experimental Section

The experimental section may have one or more chapters each containing one or more experiments. Each chapter takes the same basic form as a research article with sections for Introduction, Materials and Methods, Results, and Discussion. The arrangement of the content of these sections is, however, quite different from that found in research articles. The Introduction is very short because a great deal of the background has been given in the Review. It is sufficient to extend the arguments already made in the Review and complete them with a specific hypothesis for the experiment. Similarly, most of what would normally go into the Materials and Methods of a research paper has already been covered earlier in the Materials and Methods chapter. Only specific information, unique to the experiment, need be given and in most cases this consists of a simple statement of the experimental procedure. In fact, the words Experimental Procedure are often substituted for Materials and Methods to avoid confusion with the chapter of that name.

The results are given in full and are prepared and arranged in the same way as we have seen for a research article. Freed from the threat of an editor's red pencil, some students present results far more expansively than they should. In many cases this is simply a lack of self-discipline. Sometimes, however, it is worth recording some results which would not normally be published or publishable, but which might nonetheless be useful raw data for future workers. Analyses of feedstuffs or of composition of plants often fit this category. Rather than cluttering the main Results section, and therefore the whole report of the experiment, these analyses can be compiled in tabulated form in appendices. The appendices, containing this material from all of the experiments, are then presented at the end of the thesis. It is important to recognise that material in appendices is not part of the experimental story you are recording. If you find that you have to refer in your Discussion to an appendix it is a sure sign that you need to reorganise your data so that such material appears in the Results section.

The Discussion at the end deals with the results in relation to the specific hypothesis for that chapter. In other words the basis for discussion is, as always, the hypothesis being tested and goes no wider than this.

Where several experiments testing related hypotheses are included in one chapter it is preferable to present in sequence the Introduction, Experimental Procedure, and Results for each experiment and conclude with the one Discussion. In this way the Discussion brings together all of the experiments and unifies the chapter. The Discussion here must not ignore any of the hypotheses that introduced the experiments although one or more can be discussed at the same time if the opportunity presents itself. This form of presentation is aided greatly by the shortness of the Introduction and Materials and Methods sections for each experiment.

The General Discussion and Conclusions

At this point we return to the original unifying hypothesis and commence the Discussion based on all of the results and their support or rejection of the hypothesis. The value of a well-chosen unifying hypothesis now becomes apparent because it allows discussion and comparison of results between experiments. Until now each experiment should have been discussed separately and in isolation. Attempts to integrate one experiment with another during the course of the experimental section generally end in confusion. When results from experiments which are not yet presented are used to develop arguments the reader has trouble following this sort of gymnastics. By contrast, a complete integrating discussion in a separate final chapter can be logically arranged and is usually the most informative section of the thesis.

Whether or not you write a Conclusion segment at the end of the General Discussion is a matter of preference. Some people think that at the end of a long thesis some condensed wisdom is desirable to highlight the main points of the thesis. Others feel that this is adequately covered in a good Summary and believe that the General Discussion is so important that it should not be cluttered with such appendages. My own view is that the whole point of the General Discussion is to draw conclusions and that listing these in the Summary is sufficient.

The Bibliography

The Bibliography or References section of a thesis is no different from that of a scientific article or review except that it is generally longer. Most universities are not as stringent as editors of journals about the detailed format of references. Nonetheless, it is important that once you have decided on a format it should be followed consistently for each reference. You may wish later to rearrange material from the thesis to construct one or more articles for publication. It is wise therefore to use a format for references that includes complete titles, citation of journals, and first and last page numbers. Only in this way can you be sure of having all the material at your fingertips when you come to prepare separate articles which will meet the demands of editors.

The Summary

Where a thesis is relatively short, the Summary has the same purpose and the same form as a Summary for a scientific article. When the number of experiments, and therefore the volume of results, are large, some trimming may be necessary. Summaries of five or six pages are no longer summaries. The technique in this case is to make a list of the main conclusions that you have drawn in the course of writing the General Discussion. These will constitute the final part of your Summary. Ahead of this you then describe the principal results that led to the conclusions you have made. In this way minor results and those that do not fit the theme of the thesis are eliminated from the Summary. They still, of course, play their minor role in the body of the thesis. After this you can add, at the beginning, an abbreviated introduction consisting of little else than the unifying hypothesis. The Summary is completed with, at the end, a statement of acceptance or rejection of your unifying hypothesis, which may take the form of a final conclusion if this is appropriate.

The Table below summarises the main components of a thesis.

Table 4   The anatomy of a thesis

These are general guidelines—students should be sure to check with the guidelines at individual universities for local rules and variations.

Title Page
Index and Acknowledgements

Chapter 1—General Introduction
   The general hypothesis and a series of statements which make it a
   sensible hypothesis to test.

Chapter 2—Review of the Literature
   A review embracing all those aspects of the literature that are relevant
   to the experimental section plus extra material necessary to make a
   complete story.

Chapter 3—Materials and Methods
   All of the materials and methods common to two or more experiments
   —specifically excluding the experimental procedure for each experiment.

Chapters 4 to N—Experimental Chapters
   Each experiment or related group of experiments treated separately to
   include:
   1   A brief introduction and statement of the specific hypothesis(es).
   2   Experimental procedure.
   3   Results.
   4   Discussion of the Results in relation to the specific hypotheses.

Chapter (N + 1)—General Discussion

A discussion of the results of all the experiments in relation to the general hypothesis in the General Introduction.

Summary

1 A re-statement of the general hypothesis.
2 The overall procedure for the experiments.
3 The main results and their significance.
4 The general conclusion.

References

A careful compilation of all cited references and no others.

Getting Down to Business in Writing the Thesis— The Working Summary

The two questions asked universally by students preparing higher degree theses are:

1 (Before writing commences.) Have I sufficient research material to write up for my thesis?
2 (After writing has begun.) Where am I in this sea of data and words?

These questions arise from the sheer size and complexity of a higher degree thesis. To answer these questions we must first reduce the available material to its most essential and important elements. Once this is done we can make judgements and comparisons within and between experiments and sections of the thesis. We can think of an analysis of this kind as a working summary. The working summary vaguely resembles the Summary of the thesis but differs from it because it emphasises only those things that are vital to you, the author. By contrast, the Summary which appears in the thesis and which should not be written until most of the thesis is complete, must contain those components of methodology and of justification that you can take for granted in the early stages of writing.

To construct the working summary, we begin by taking out the vital elements from the experimental section of the thesis. These are:

1 The hypothesis, or hypotheses.
2 The main results (preferably in order of importance).

3 The main discussion points arising from the results (also in order of importance).

This information should be carefully extracted from each experiment to be presented in the thesis.

As an example, let us assume that the study of the relationship between plant associations and the pathological fungus which we looked at previously (on page 105) has been completed and takes the form of a series of experiments. The working summary, which can be in an abbreviated form because only the student and his supervisor need to understand it might include a section like this:

Experiment 6

HYPOTHESIS:
That the exudates from indigenous species of *Leguminosae* restrict the growth of the pathogenic fungus *Phytophthera.*

MAIN RESULTS:
1 (Experiment 1). Counts of *Phytophthera* were lower in the soil taken from the root zones of leguminous plants than from the root zones of other plants.
2 (Experiment 2). Culture plates of *Phytophthera* were inhibited when live root tissue of legumes was added but not when dead root tissue was added.

MAIN CONCLUSIONS:
1 Hypothesis supported in each experiment.
2 Inhibitory substance is only found in living tissue. Therefore a new hypothesis: that under field conditions leguminous plants must be actively growing to inhibit the fungus.

Experiment 7

HYPOTHESIS: (As above)
MAIN RESULTS: etc.
MAIN CONCLUSIONS: etc.

If we then add the general hypothesis from the Introduction section we can use this hypothesis as the basis for developing the General

Discussion section from the summary of the Results and Discussion topics of individual experiments. By now it will be apparent to both student and supervisor whether or not there are gross deficiencies in the group of experiments. This form of summary should also suggest what further experiments need to be done to complete a coherent series that will result in a worthwhile thesis.

Reducing the experiments to their essential elements in this way may seem simple but in practice it can be very difficult because these few statements are the result of a great deal of the original thinking and analysis that will go to make up the thesis. It is not unusual for a working summary of this kind which may be only three or four pages long to take a month or more to construct. Once the summary is complete, however, the writing becomes a matter of filling out the details and, with the summary close by, it is virtually impossible to become lost in the large mass of material that will go to make up the bulk of the thesis. The supervisors who need to read and comment on drafts of sections of the thesis can do so sensibly and with confidence if they too have a copy of the working summary beside them to allow them to appreciate the perspective of the section that they are reading.

Using the working summary

With a carefully planned working summary before us we can now begin the detailed writing.

- Each Introduction will be a justification of the hypothesis or hypotheses proposed for each section.
- Each Experimental Procedure will be an outline of the experiments to test these hypotheses.
- Each Results section will be written so that the main results specified in the working summary will be emphasised. Tables, graphs, and text will all be drawn up with these main results in mind. Other, less important, results will also be included, but their position and mass should indicate their relative lack of importance.
- Each Discussion section will also be constructed from the working summary using the principles we have already covered in Chapter 1 for developing discussion topics.

The main purpose of the Review of the Literature is to provide background for, and to introduce, the hypotheses. The working summary is useful here, too, as a form of checklist which can be used as a framework for the Review of the Literature.

It would be surprising if, during the writing of the thesis, new ideas did not emerge. Such ideas can be incorporated into the working summary without reducing its effectiveness as an outline for the complete thesis. On the contrary, the working summary will assist you to weave new ideas into the fabric of the thesis by suggesting exactly where they should be included.

If we consider the working summary as the first draft of the thesis, then the expanded version we have just developed can be considered as the second draft. At this stage a student writing a thesis should take advantage of having a supervisor who is, or should be, an officially appointed and readily available participant for the 'colleague test'. A supervisor's experience in writing papers and in supervision of other students, will be invaluable at this stage. Nevertheless, be careful that their familiarity with the work does not result in their missing badly phrased expressions and jargon. If you can obtain it, a second opinion, even on selected sections of the thesis, may be very helpful as a guide to the readability of your work. Make sure that anyone whom you have induced to read sections of the thesis has access to your working summary so that they know where they are.

Many students find the writing of a thesis tedious and consider it an inordinate waste of time. In case you are tempted to think similarly, remember the principle expressed repeatedly throughout this book: good writing and good science go hand in hand. The training you are undertaking when writing your thesis is every bit as important as your research work. Treat it as such and your skill as a scientist will be enhanced. Your immediate colleagues may recognise you as a fine scientist and a thoroughly nice person through personal contact but your standing with the other 99.99 per cent of the scientific world will depend upon how well you write.

9

Writing science for non-scientists

WE ARE ALL aware of the exponential increase in the number of scientific articles written each year by scientists for their scientific peers. The rigorous peer-review system and editorial scrutiny ensure that a high proportion of scientific articles for scientists are of a high quality. There are even more articles on science or with a strong scientific content written each year for readers who have no background or interest in science.

Unfortunately, of those articles written for non-scientists, a disappointingly high proportion are of very poor quality.

There are probably two underlying reasons for this. First, if they are written by non-scientists they are often grossly inaccurate. Writers who may not always be competent to distinguish what is scientifically reasonable, may nevertheless have well-trained journalistic instincts to see 'a story'. So sensationalism overrides the truth. To exacerbate the problem many scientists when asked to supply information to journalists with little feeling for science, become very defensive because they have experienced being misquoted before in a way that may have embarrassed or offended them.

Second, if they are written by scientists, the struggle for scientific justification and exactitude, particularly in details of marginal interest, often detracts from the main thrust of the story. Many scientists find it hard to judge from the material in front of them what is likely to keep a reader interested.

In both of these cases the real value of the work may never emerge.

This need not happen. Some of the most riveting scientific stories imaginable are derived from discoveries made in pure and applied science.

Their success as stories depends on their translation from the scientific to the popular literature in a clear, accurate and appealing form.

Well-trained scientists know what they seek when they pick up a scientific article. They want hypotheses, methods, results and the cut and thrust of a good discussion. They are used to the structure of a scientific paper and they know precisely where they should seek the sort of information they require. That is why they read scientific papers. When non-scientists try to read the same article, they have no professional background to appreciate its structure nor, in many cases, its content.

So why do non-scientists bother to read about science or matters related to science? The reason is simple; science has an impact on almost everybody's lives, their work and their interests. And herein lies the key to making a successful transition between the world of the scientist and the world of the non-scientist through the written word. You must identify what is likely to attract their interest.

Why don't scientists talk a language that non-scientists understand?

Scientists, who can usually talk to one another without problem, often feel very uncomfortable when explaining themselves to people with no scientific background. There is a logical reason for this and understanding this reason is a great help in bridging the communication gap between the two.

When you ask scientists what their work is about they will usually begin by telling you in detail how they go about their daily tasks—their methodology. This is perfectly understandable because it is what occupies most of their working day. Then, they will probably quote some of their most recent or most exciting results to you. This, too, is understandable because results are the things that keep scientists motivated. Then, they may get round to telling you why they are doing the research in the first place—their hypotheses and expectations. Many scientists find this a more difficult subject to broach than the other two. It is only rarely and only after careful questioning that you will learn where they think that the research fits into the bigger picture of the scientist's discipline. Even more rarely will you find out what the research that is being done means for humanity as a whole, or at least that part of humanity to which the listener may belong. In short, it becomes more and more difficult to glean information, the further the information is from the scientist's everyday mental pathway.

On the other hand, if you ask non-scientists what most interests them about a particular scientific subject you get an entirely different set of answers. The most compelling thing that they want to learn is what is in the work that may affect them. That is, after all, a reasonable motive. Then they may seek to understand where the scientist's work fits into what they already understand about science. Then, they may want to know why the scientist is bothering to work in the field that he or she is in. After they have satisfied themselves about these three important pieces of information they may begin to show an interest in specific results and, even more rarely, in the methodology.

Here then is a most interesting phenomenon. The non-scientist seeks information in precisely the reverse order to that in which the scientist has been prepared to give it. So, scientists have problems explaining themselves to non-scientists unless they deliberately set out to alter their natural pattern of presentation.

If scientists wish to be understood, they must be careful to talk about their work under the following categories, in decreasing order of importance:

1 what is in it for the reader;
2 where it fits into the broader pattern of science;
3 why the work was done;
4 the major results; and
5 some methodology.

Only if they adhere reasonably rigidly to this pattern and forsake the 'natural' scientific order can scientists expect to find an attentive audience among non-scientists.

Attracting the reader

If you are going to write an article for non-scientists your major task is to find what is likely to interest the reader because, as with all good written material, your article must place the reader's wants and interests as top priority. The reader has to be enticed to plunge into reading and then be held until the article has unfolded.

Among the many things that can attract a lay-reader are:

Subject matter   Some subjects are universally interesting and topical. Things that save money, things that entertain or pertain to sport, things that concern—like global warming, and new possibilities for the prevention of diseases.

Timing   Some topics can be very appealing if they come before the reader at critical times. An article on a new form of thermal underwear will be more attractive in winter than in summer, as will an article on a cure for the cold. An article on a new wheat variety is more often relevant just before the sowing season than at harvest. An article on the management of rams will be read with more interest just before the mating season than during lambing.

Presentation of science at a human level   Readers sometimes like to hear about the human side of scientific discovery. The joys of a break-through, the agony of a near miss, the hard work in a back room.

Curiosity   Often readers are attracted to a new approach that science brings to a common problem, or to an apparent answer to some-thing that was hitherto a mystery or an explanation for an everyday phenomenon.

The length of the article   Whether or not an article is read often depends on its length and when and where it is likely to be read. For example, a long article might be appropriate for a Sunday magazine, but would be avoided by commuters scanning their morning newspaper. The article will usually not be successful unless it can be read at one sitting.

If the article is for a national newspaper at least one of these ideas can give you a basis to build the story around. But it is often even easier than this to find an 'angle'. The readership is seldom going to be totally heterogeneous—but it is likely that they will have a common background or interest. The article may be for an industrial journal, a rural journal, a gardening magazine, or for any number of prescribed single-activity audiences. In this case the focus and the interest can be very easy to identify. The important thing is not to start before you have identified it.

The essential ingredients

Your article for the non-scientist is a concentrated piece of writing. It must contain the three ingredients of scientific writing; precision, clarity and brevity.

Precision in this case is not necessarily associated with large masses of figures, but more with conveying what the scientists who did the work feel is an accurate description of what they have done and what they have proposed. The only way to assure yourself of this is to allow the scientist or scientists whose work is being described to read and to be happy with the last draft. The last draft is the one that will be printed, not an intermediate one that might change after the scientist has seen and approved it. Writers who gain a reputation for consistently adhering to this procedure gain the confidence of scientists and gain access to material that would otherwise remain unavailable. You may have to modify an enthusiastic, but inaccurate, impact to earn that confidence, but you will establish a record of credibility with both the scientist and the reader that in the long run is both sustainable and praiseworthy.

Clarity is always important but in articles for non-scientific readers it is often the main reason for writing the article in the first place: to clarify and make accessible for a non-scientist what is presumably comprehensible for a scientist. Needless to say, scientific terms that may be confusing must be transposed to everyday language. Concepts that are well accepted by scientists may need to be spelled out more fully or illustrated with examples for non-scientists. It is essential to do this very carefully because, with this type of article, readers seldom give you a second chance if you confuse them .

The length of the article is often prescribed. If it is meant to occupy a page, or half a page, or 500, or 1500 words it usually means precisely that. So, brevity too, is a constant obligation. Generally, when you begin to write you will have more to say than you have words or space to say it. To achieve a good result you have to decide what you can afford to cull from the material in front of you and then be economical in the way you express what remains. If 500 words means exactly 500 words, you could find yourself having to delete just one or two sentences or to manipulate the structure of the others to meet your target precisely.

Constructing the article

A well designed article has four main components:

The title
The summary
The description
The follow-up

The title

In this context the title is often a headline rather than a title. It is very short and destined to hit a key nerve in the reader. Needless to say the choice of words in the title is a paramount consideration. It has the dual role of catching the eye and providing a strong indication of the contents of the article. Nonetheless, be careful. Joke books abound with 'howlers' taken from newspapers and other magazines in which the headline presumably made sense to the writer, but made no sense, or a different sense, to the reader. 'Justice split plan sparks probe fear' boldly announced a national newspaper—hardly enlightening stuff. Neither is 'Reds beat blues—try the difference' on the sporting page.

The summary (or header)

Normally the summary is not entitled 'summary' as it is in a scientific article, but it fits the same role and provides the reader with a short, succinct version of the whole article, complete with the punch line, the take home message, and any wisdom that the article attempts to get across. Quite often the header/summary is distinguished from the rest of the text by having a larger font size or a different disposition on the page from the main text. For example, it may span two normal columns of ordinary text.

The description (or detail)

You will see already that one of the main principles of scientific writing, that there are no secrets and no build-ups to great revelations, applies equally in this form of scientific writing. You said what the article is about in the title, you reiterated it and expanded on it in the header/summary and now, in the body of the text, you present details for those whose interest is still strong enough and who have enough time to read on.

But this is not the stage to become careless. The detail that a scientist must seek is not necessarily that which your non-scientific reader wants. Keep the reader in mind to the end. And keep to the principles of fluency and reader expectation discussed earlier regarding the style of a scientific article. Above all, non-scientists are not committed to read on if they become confused. They read something else.

Follow up

If you have done a good job, at least a proportion of your readers will want to know more. A good article on a scientific theme will conclude by directing them to a more detailed article, a companion article, or to the scientist or laboratory where they can get more information.

The final inspection

When you have finished writing and have checked, as usual, for typographical, grammatical and spelling errors check also that the article is suitable for its readers. This can be done simply by ensuring that it meets just five important criteria. It must be:

1 attractive and fresh to get the attention of the reader in the first place;
2 bright and relevant to the reader;
3 informative, not only in its own content, but in its directing the reader towards further information;
4 accurate in what it says and feasible in what it claims; and
5 the right length.

A friend of mine has created a career in translating scientific material into exciting reading for lay audiences by following these simple rules. It has become such a challenge that she makes a feature of taking erudite and seemingly unlikely scientific articles and constructing them into attractive scientific stories. The results are spectacular and her most satisfying approbation comes from the scientists whose work she has been describing who have failed to realise until they read her interpretation how potentially popular their work can be.

Index